Flowerdew Hundred

*The Archaeology of a
Virginia Plantation,
1619–1864*

Flowerdew Hundred

The Archaeology of a

Virginia Plantation,

1619–1864

James Deetz

UNIVERSITY PRESS OF VIRGINIA

Charlottesville and London

The University Press of Virginia
Copyright © 1993 by the Rector and Visitors
of the University of Virginia

First published 1993
First paperback edition 1995

Library of Congress Cataloging-in-Publication Data
Deetz, James.
   Flowerdew Hundred : the archaeology of a Virginia plantation,
1619-1864 / James Deetz.
      p.   cm.
   Includes index.
   ISBN 0-8139-1461-2 (cloth). ISBN 0-8139-1639-9 (paper).
   1. Flowerdew Hundred Plantation Site (Va.) 2. Excavations
(Archaeology)—Virginia—Prince George County. 3. Plantations—
Virginia—Prince George County. 4. Prince George County (Va.)—
Antiquities. I. Title.
F234.F56D43   1993
975.5'585—dc20   93-9226
                                                              CIP

Printed in the United States of America

To

David Harrison III who made it possible

and to

Pinky Harrington who made it work

Contents

Illustrations

Tables

Acknowledgments

THIS BOOK is the product of the efforts of literally hundreds of people, covering nearly a quarter of a century. Archaeological research is always a team effort, for the careful removal of the earth in which the remains of past settlements are contained is a slow and often tedious pursuit, and numbers of field crew members are needed if any significant amount of excavation is to be accomplished. Such has certainly been the case on Flowerdew Hundred Farm, on the south side of the James River, in Prince George County, Virginia. Here excavators ranging in age from early teens past retirement have put up with the heat, humidity, chiggers, mosquitoes, and other blessings of the Virginia summer and recovered the thousands of objects and exposed the house foundations, fortifications, wells, and trash pits on which this work is based. The results of this labor are nothing less than spectacular, and it is time to assemble all of the evidence and construct a story of the growth of American society as it has been witnessed on a single plantation in the Virginia tidewater.

From 1971 through 1978 the excavations were carried out by the College of William and Mary, under the direction of Professor Norman Barka, assisted by field director Leverett Gregory, Charles Hodges, and Andrew Edwards. In 1980 I initiated a research program, and it has continued to the present. These excavations have been supported in part by the University Research Expedition Program, University of California, Berkeley, the National Endowment for the Humanities, and University of California field school students. During the entire twenty-one years the generosity of the owner of Flowerdew Hundred Farm, David A. Harrison III both as gracious host and benefactor, has made it possible for the program to continue and thrive. Mr. Harrison has provided more support for archaeology than has any other private individual in the state, and this book could never have been written without his abiding interest in the story of our collective beginnings, and where those have led us as a nation.

To date eleven sites have been excavated, in part or completely, under the direction of no fewer than seventeen field archaeologists of various affiliations. Rather than clutter the pages that follow with names, dates, and site numbers, it seems better to provide this important information in one place, so that the reader might easily refer to it should the need arise. Sites 64 and 65 have seen the longest ongoing excavation, and further investigation of them has yet to be done. Norman Barka, Charles Hodges, Andrew Edwards, and Leverett Gregory worked at both of these sites between 1971 and 1978. Further work was carried out under the direction of Leonard Winter, a University of California graduate student from 1981 through 1984. During the summers of 1988 and 1989, additional excavations were directed by two Flowerdew Hundred staff archaeologists, Scott Speedy and Ann Markell, and a University of California graduate student, Margot Winer. Site 66 was investigated during the 1973 and 1974 field seasons, under the direction of Gregory, who also excavated site 72 during the same period. Site 77 was studied by another University of California graduate student, Matthew Emerson, and formed the basis of his Ph.D. dissertation. Site 82 was excavated in part by Hodges and Taft Kiser, both Flowerdew Hundred staff archaeologists at the time. Another Flowerdew Hundred Archaeologist, Ann Markell, directed the work at site 92 during the seasons from 1985 through 1989, and she also prepared her University of California doctoral dissertation on the site, as well as writing a detailed site report. Site 97 (now subsumed under the number 98) was first excavated by Mary Beaudry with a William and Mary field school group in the summer of 1978. Further work was done at the site from 1984 through 1986, under the direction of archaeologists from the Robert H. Lowie Museum of the University of California, Robert Wharton and David Herod. Between 1986 and 1992 site 98 was investigated by a number of workers, including Kiser, Wharton, Speedy, Markell, and myself. Finally, sites 113 and 114 were studied by Margaret Scully (1980–1983) and Lawrence McKee (1984–1986), respectively. Both were graduate students at the University of California, Berkeley, and they prepared Ph.D. dissertations on the sites.

Little of what follows is the result of excavations I directed, save in the most general of ways, as overall project director. I directed the first season's work at site number 113, a brief program of testing lasting only two weeks, and most recently, the excavations at site 98, assisted by a California graduate student, Maria Franklin. Several extended discussions of the archaeology at Flowerdew Hundred in this book are based on, and indeed paraphrase, the work of others, named above. This is particularly true in the case of Ann Markell's interpretation of the evidence at site 92 and Larry McKee's study of slave foodways as seen at site 114. Furthermore, the lengthy treatment of the "Colono ware problem" owes a tremendous debt to the pioneering work of Leland Ferguson of the University of South Carolina, whose recent book *Uncommon Ground* was of critical importance to the interpretations put forth here.

Since I suffer from a serious case of computer phobia, it was necessary for other willing people to deal with the tremendous task of transferring a poorly typed (two fingers) manuscript to disk. Gail Taylor and Bob Wharton managed to accomplish this not inconsequential feat through the first four chapters, and Trish Scott, Avis Worthington, and Jim Allan completed the job. In addition, were it not for Ms. Scott's monumental job of editing the entire manuscript, even what was transferred to disk would have been a copy editor's nightmare. Appreciation also must be expressed to those who read various versions of the manuscript and commented, especially Marley Brown, Martin Hall, Pinky and Virginia Harrington, David Harrison, Taft Kiser, Ivor Noël Hume, Tonia Rock, Trish Scott, Gail Taylor, and Bob Wharton. Alison Bell prepared the index and assisted in reading page proofs.

Many people helped in the production of the book. At Colonial Williamsburg Rob Hunter of the Collections Department provided hours of assistance in selecting and arranging for the photographs of ceramics in the appendix and further help was provided by Bill Pitman of the Archaeology Department and Catherine H. Grosfils of the Audiovisual Library. Except where noted, the drawings were produced by my son, Eric Deetz, and the photographs by Eugene Prince of the Phoebe Hearst Museum of the University of California, and Margot Winer provided her original drawing of the Hall, in Salem, South Africa. At the University Press of Virginia, Director Nancy Essig was always encouraging and Carol Rossi had the unenviable task of keeping me on schedule with grace and gentle prodding. A portion of chapter 1 is reprinted by permission of the publisher from my article "American Historical Archaeology: Methods and Results," *Science* 239 (Jan. 1988): 362–67, © AAAS.

The day-to-day support of the entire Flowerdew Hundred Foundation staff during field programs was indispensable to their success. Earlier, Director Tom Young and, later, Executive Director Bob Wharton were always on hand to deal with a myriad of problems, from broken plumbing to storm-torn tents, and Gail Taylor kept everyone and everything on track. Jan Stenette was most helpful in digging out various obscure references. Farm Manager Tommy Banks was always willing to lend his equipment and expertise in removing massive quantities of soil from one place to another, relocating spoil heaps, and stripping the topsoil whenever the need arose, saving literally thousands of hours of work, had it been done by hand. And last but not least, the people at Parker's Store in Garysville always made us feel especially welcome and provided everything from country hams to cold beer.

Flowerdew Hundred

The Archaeology of a
Virginia Plantation,
1619–1864

Chapter One

July 27, 1626, Flowerdew (Peirsey's) Hundred, Colony of Virginia

A LICE THOROWDEN arose before dawn, cross and weary. It had been a bad night, first the mosquitoes buzzing and biting, and then the thunderstorm that rolled down the valley, driving the river hard against the shore, tearing branches from trees and leaving great puddles which would be churned into a quagmire when wagons were driven through them. Yesterday had been trying, too much to do and not enough time to do it. She and Katherine, the other maidservant, did their best, she thought, but since himself had built this new house, there seemed no end to the chores—clean this, fetch that, mend this—on and on it went. Still, Alice allowed that she and Katherine should be thankful, for their life in London had been much worse, and when they boarded the *Southampton* in 1623, it was with high hopes. But Virginia was such a strange and frightening place. Bugs of all sizes, shapes, and colors were everywhere, and the night echoed with sounds such as she had never imagined, much less heard. Far worse, only four years ago, the Indians had killed hundreds of people with hardly a by-your-leave. People had been falling sick lately from who knows what, and just yesterday a man and an infant had died and would be buried today in the master's yard, them being favorites of his. But the house was a grand one, with its huge brick chimney and fireplaces in both upper chambers, and she and Katherine had snug quarters in the loft. But still . . .

Just beyond the garden fence, Alice could hear two of those strange black people speaking softly in a language she would never understand, although in truth, she had little better luck understanding Henry Carman, who hailed from Cornwall. It was barely light, but the air was heavy and warm, and Alice knew it would be another hot day when the sun would appear, a bloodred ball just above the trees. Putting her cares aside for the moment, she took her pipe from beneath her apron, packed it with tobacco, and entered the hall. Reaching into the embers of last night's fire with a small pair of tongs, she took out a glowing coal and began to light her first pipe of the day. The ember flared and sputtered, showering sparks over Alice's hand. She dropped the pipe, and it fell to the floor, shattering to pieces.

When Alice broke her smoking pipe, as one might suppose while working with the fragments, she little dreamed that three centuries later those fragments would be used to bring her world back to life, to give people in the far distant future some sense of what life was like in the seventeenth century. The pieces were simply swept aside, to find their way eventually into a pit with other refuse—the remains of a meal, pieces of a broken jug, and some old useless scraps of metal. But to archaeologists of the twentieth century, the fragments of pipe stem, looking for all the world like bits of blackboard chalk with holes drilled through them lengthwise, provide a key to such understandings. To know how, and why, we must place Alice in her little corner of the world, and her broken pipe with the other realia of her time.

Any story must have a beginning, although one can choose from a number of possibilities. For our account, a plain little pipe stem fragment seems particularly appropriate, for it is possible to build on such a simple uncomplicated object to reach understandings of sweeping changes in the way people viewed one another and the world in which they lived, of the great transformation of American society from medieval to modern. But first, the pipe stem fragment; it can be thoroughly described in a very few words. Length, an inch and a half, an accident of breakage; diameter one-third inch; diameter of the hole that passed the smoke, $8/64$ inch. It is made from white clay, technically called ball clay, fired quite hard, and shows some slight burnishing. It was made in England and made its way across the Atlantic to Alice at her residence in the Chesapeake colony of Virginia, where she was serving her seven-year indenture to a planter at Flowerdew Hundred, a plantation on the south side of the James River. This was Alice's world, where she would live out the rest of her days. She and hundreds of other people would unintentionally leave behind a record of their passing, a record of things, not words, to be recovered by archaeologists, allowing them to piece together an exciting story of the growth of English culture transplanted to a strange new world.

In an area rich in history, Flowerdew Hundred occupies a special niche. Founded in 1619 by George Yeardley, Virginia's first royal governor, it witnessed cameo appearances by great and not so great figures in American history. Fifteen of the first twenty black Africans to come to the English colonies resided at Flowerdew Hundred. It survived the attack on the colony by the Powhatan Indians under Opechancanough in 1622, with a loss of only six lives. It was along Flowerdew Hundred's river shore that gunboats under the command of Benedict Arnold shelled buildings on the plantation. Edmund Ruffin, a local resident, almost certainly visited Flowerdew in the early nineteenth century before becoming famous for supposedly firing the opening shot of the Civil War at Fort Sumter. In 1864 the entire Union Army of the Potomac, under the command of General Ulysses S. Grant, crossed the James River at Flowerdew Hundred in an attempt to outflank Robert E. Lee's Army of Northern Virginia. Union troops bivouacked at a house on the plantation before moving on to lay siege to Petersburg in the closing months of the conflict. Between these moments of high drama, hundreds of people made Flowerdew Hundred home, pursuing their daily routines of farming the land. The places where they lived and worked are known today, the result of painstaking archaeological site survey of the property, walking plowed fields to identify the telltale evidence of past human presence—bits of broken pottery, pipe stem fragments, animal bone and oyster shells, and small pieces of what once were houses, including nails, brick chips, window glass shards, and mortar. It is these small scraps of a once vibrant and dynamic world that provide us with our initial access to time past.

Carefully collected, washed to rid them of clinging soil, and numbered in ink to document the place from which they came, they are spread out on a tabletop and sorted, classified, and ordered into groups that with luck might reflect various aspects of the lives of the people for whom they were the very stuff of life. Pieces of colorful pottery not only tell us how food was prepared and consumed but also provide us with at least an approximation of the time that they were in use, since we have a good knowledge of the history of the ceramic industry in Europe from which so much of the pottery came. Bits of brick, mortar, window glass, and nails provide hints of the kind of house that may have once stood on the site. And the pipe stems provide us with a way to organize all of the sites into a wider picture, done first in broad strokes and then filled in and detailed through intensive excavation of the various key sites in question. So we must first enter the esoteric realm of the clay smoking pipe, to see just how and why this is so.

Tobacco was introduced into English society in the late sixteenth century, and the preferred way of taking tobacco smoke was through a pipe, loosely modeled on those encountered by English colonists in North America. In contrast, the first Spanish contact with tobacco was in the West Indies, where cigars were the chosen way of smoking the plant, and so the cigar became typical of the Hispanic colonial world. Pipe manufacture sprang up quickly in England, and by the time of the settlement of Jamestown in 1607, large numbers of craftsmen were producing pipes to meet the needs of a population besotted by the weed. In spite of King James's invectives against tobacco, its popularity continued to grow, and the first colonists to come to Virginia were already addicted consumers.

The earliest pipes were very small by later standards, with bowls shaped rather like an acorn, holding only a small amount of tobacco. This may have been due to the high price of tobacco in the early years of its use; in time the bowls became larger and the stems longer, although there was always considerable variation in stem length. These two changes in the shape and size of pipes had an important effect on another part of the pipe, the hole through which the smoke passed from the bowl through the stem. As time passed, the diameter of the stem bore gradually was made smaller, apparently at a rather constant rate. It is not clear just why this is so, but the most probable explanation relates to the way the pipes were made. A two-piece mold was employed in which a piece of green clay called a tadpole was placed. After the mold was closed on the clay, the bowl was hollowed out and the stem bored by passing a wire through the stem from mouthpiece to bowl. It seems likely that the lengthening of the stems over time had two effects, either of which would require a smaller bore. First, larger bowls would hold more tobacco, and as a result, a smoke would take longer if all the tobacco was consumed. Indeed, it would seem that users of tobacco in the earliest years drew smoke in quickly through the large-bore stems and consumed a bowlful in a relatively small number of gulps. This way of taking tobacco was referred to as "drinking," almost certainly due to its quick wholesale consumption. But bigger bowls meant longer smokes, and anyone who has used a clay pipe knows how hot it becomes, well up the stem from the bowl. So, longer stems were desirable. But longer stems require more precision in the passage of the wire, and a smaller diameter would facilitate such precision. Then, too, a smaller bore would allow less smoke to pass through, an equally desirable result, given larger bowls and more tobacco per smoke. But no matter what the exact reason, stem bores were reduced in size by pipe manufacturers as time passed, but this change was lost to human knowledge until it was rediscovered by Pinky Harrington in the early 1950s.[1]

Jean C. Harrington, known as Pinky to his colleagues and friends, is acknowledged generally as the Father of American Historical Archaeology. In the course of his work on the analysis of material excavated from the site of Jamestown, he did what any archaeologist working on seventeenth-century English colonial sites would have done, identify and date the many pipe bowls in the collections. This task was relatively easy, for by the 1950s there was a considerable body of careful scholarship on the evolution of the English clay smoking pipe. This research concerned itself exclusively with pipe bowls, for their shape and size could be shown to change in a clear progression (fig. 1). Furthermore, makers' marks were stamped on many of the bowls, and these had been identified as to individuals whose dates of pipe manufacture could be determined from documents such as directories and apprentice rolls. But the problem with pipe bowls was that there were not very many of them, compared to the great number of stem fragments from any given site. So few datable bowls, and so many stems, of no apparent value save to give some sense of the amount of smoking that might have taken place—a common archaeological dilemma wherein the commonest objects seem often to provide the least amount of useful information. But when Pinky Harrington noticed that earlier pipe bowls had larger stem bores the stem fragments suddenly became important tools in determining the date of the sites from which they came. This was serendipity in its purest and most wonderful form, the discovery of something useful while searching for something else.

It became apparent that the progressive reduction in bore diameter over time proceeded at a reasonably constant rate, so all one need do when deriving a date from stem fragments at a site is to compute the percentage of stem bits of each bore diameter and plot them in the form of simple histograms. Harrington measured the diameters of his stem fragments using a set of six drill bits ranging in size from $9/64$ through $4/64$ inch. The time segments for the six bore diameters are:

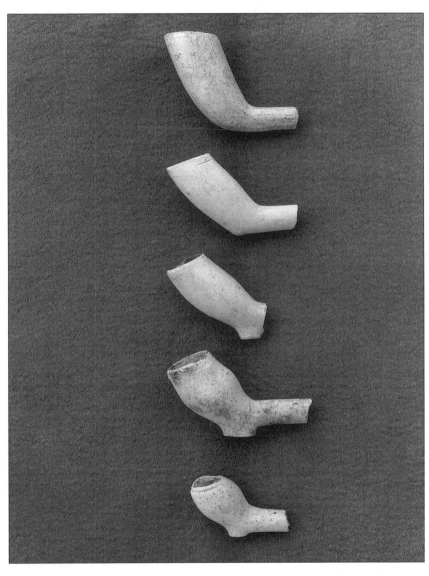

1. Changes in tobacco pipe bowl shape: early (*bottom*) to late (*top*). Approximate dates (*bottom to top*): 1600–1640, 1620–60, 1650–80, 1680–1710, 1720–1820

Inch	Dates	Inch	Dates
$9/64$	1590–1620	$6/64$	1680–1710
$8/64$	1620–1650	$5/64$	1710–1750
$7/64$	1650–1680	$4/64$	1750–1800

A site which produces stem fragments with bore diameters of $7/64$ inch and a few of $8/64$ and $6/64$ inch probably was occupied between 1650 and 1680. Stem fragments from sites occupied for periods of time longer than thirty years or so will show a wider range of bore diameters and fewer of each size, a fact that will assume considerable significance in our Flowerdew Hundred story, for the hundreds of stem fragments picked up from the surface of eighteen sites in the James River bottomlands held another secret that was to be revealed through yet another serendipitous happening.

On a chilly afternoon in April 1984, archaeologists working at Flowerdew Hundred, acting out of boredom as much as anything else, engaged in a little exercise in site comparison. The site survey records had recently been updated and cast in a different format. The new forms had spaces for a range of relevant data—location, key artifacts, pottery types, preservation—and, of course, a summary of the pipe stem bore diameters for each site. It seemed a good idea to produce Harrington-type histograms of the bore diameters for each site on clear acetate, so that any two sites could be compared simply by placing their histograms together, a very low-tech device but effective nonetheless. Over a lunch of Stroh's beer and Prince George County surf and turf (sardines and Vienna sausages), someone got the idea of stacking all eighteen histograms atop one another. No one was looking for anything special; it just seemed a good idea for want of anything else to do at the moment. Had anyone predicted a result, it probably would have been a colorfully chaotic mass of lines, since each site's graph was produced in a different color of ink. But by the time the seventh or eighth acetate was added to the stack, a pattern began to emerge, and when all eighteen sheets were piled up, the result was clear and unmistakable. Rather than showing a fairly even distribution of sites across time, the histograms fell into three discrete groups, quite separate one from another. These aligned histograms were combined by totaling all stem fragments from the sites in each group, producing three aggregate graphs (fig. 2).

The seven sites in the earliest group all showed a sharp peak of occupation during the period from ca. 1620 through 1650, with large numbers

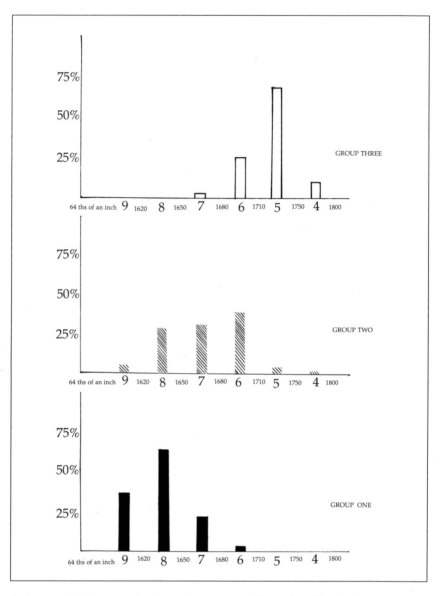

2. Aggregate histograms for three site groups at Flowerdew Hundred

of stem fragments with $\frac{8}{64}$-inch bores. The number of $\frac{7}{64}$-inch bores drop off sharply, indicating that all seven sites were probably abandoned during the next thirty years. Where the people went, we do not know; the only fact that is certain is that they were no longer living at these sites. The second, intermediate, set of sites, six in number, have a very different pattern of stem bore distribution, exhibiting a much flatter graph with bores ranging in diameter from $\frac{9}{64}$ through $\frac{4}{64}$ inch and a peak in the later seventeenth century. The latest set of sites, numbering five, shows a peak as sharp as the first set, this time at the $\frac{5}{64}$-inch stem bore point, indicating they were occupied during the first half of the eighteenth century. Taken together then, the eighteen sites suggest an initial buildup of occupation at Flowerdew Hundred, dropping off in the second half of the seventeenth century, and another similar intensification of occupation during the first half of the eighteenth century. But some people remained on the plantation over a longer period of time, as indicated by the intermediate group.

The people who once lived on all of these sites were subject to the effect of events taking place far beyond their immediate horizon. While living three-thousand-odd miles from their English homeland, they were far from isolated from the world on the opposite side of the Atlantic. Objects from Germany, Holland, Spain, and Italy as well as from England are found in the ground at Flowerdew and serve to underscore this fact. For this reason, it is quite likely that a proper understanding of the pattern of occupation indicated by the pipe stem analysis can be found only by directing our attention beyond the James River and beyond the Virginia colony to the English colonial world as a whole. Two things stand out as particularly powerful forces that shaped the lives of colonial Virginians— tobacco and the labor to produce it. Until John Rolfe, the future husband of Pocahontas, introduced a highly prized variety of tobacco from Trinidad, Virginia's future looked grim at best. Settled by a rather gentrified class of Englishmen, who had high hopes of gold and other riches, the colony went through several disastrous years, culminating in the starving time of 1610, when people resorted to cannibalism to survive. But the new tobacco variety changed all of this, and the years between 1619 and 1670 were true boom times. Get rich and get out seemed to be the byword, though far more people died of disease than succeeded in making a quick fortune on England's newfound burning need for tobacco. The boom lasted five decades, to be ended by a sharp decline in tobacco prices.

Tobacco is a labor-intensive crop. From the start land in Virginia was

cheap, but labor, expensive. For a while labor needs were met largely
through indentured servitude, but indentured servants were not exactly
ideal for the purpose. A much better supply of labor was to be found on
the African continent, and England became involved in the slave trade
almost from the beginning. There were never more than a few hundred
Africans in the colony from 1619 until the 1680s, but during that decade
the number of slaves increased dramatically, to reach over six thousand
by the opening years of the eighteenth century. So it is that the history
of seventeenth-century Virginia is enclosed in a kind of parentheses, the
tobacco boom on one side and wholesale, fully institutionalized slavery
on the other. These two events coincide neatly with the first and last periods of settlement at Flowerdew Hundred and provide us with at least a
starting point from which to begin to fit the archaeology of the plantation
to the history of the colony and the world of which it was a part. Such
an exercise is known as a "research design," that is, a somewhat general
model that will allow one to fit all of the parts into a coherent whole, with
little or nothing left over to be explained in other terms. At Flowerdew the
fit is indeed comfortable, and the accommodation would not have been
possible unless all of those pipe stem fragments had been carefully measured, the histograms developed, and the historical record consulted. But
no matter how neat the fit between archaeological data and historical
fact, we must not lose sight of the central matter in all of this. Eighteen
sites mean at least eighteen families, as many as two hundred people.
Masters, their wives and children, indentured servants, and slaves all lived
at Flowerdew, where they were subject to great historical forces which
shaped and directed their lives, and all left some trace of their passing in
the ground.

Although the elements, the feller's ax, and the farmer's plow have
erased and dispersed many of these traces, there are still enough remaining to tell a story worthy of note, and archaeology plays a central
role in its construction. Historical archaeology in America has come of
age during the twenty years following its emergence as a separate and
distinct subdiscipline when the Society for Historical Archaeology was
founded in 1967. Since then, there has been a vigorous development of
archaeological approaches to the accounting of the American experience
since 1492. Any archaeology that deals with the material remains of literate people is historical; thus the archaeology of ancient Greece and Rome,
ancient Sumer, or dynastic China is historical since these civilizations
were fully literate and left ample documents. But in practice historical

archaeology as it is conducted in the United States is usually restricted to the study of European Americans or other people whose presence resulted from European settlement—African Americans and Asian Americans—and of the native Americans in the years following initial European contact as they interacted with the new arrivals from the Old World.

One common definition of historical archaeology is "the archaeology of the spread of European culture throughout the world since the fifteenth century, and its impact on indigenous peoples."[2] This is a post hoc definition, describing the work of most of those who consider themselves historical archaeologists. The study of southern plantation life, slave and free black communities, Chinese labor camps, New England Puritan farmsteads, frontier forts, French fur-trading posts, Spanish missions, and historic Indian pueblos all fall within this definition, to cite but a few examples. In every case historical archaeology attempts to ask sophisticated questions of its data, couched in terms of modern historiographic and archaeological methods. It has not always been so. The field has a long history; as early as 1856 James Hall, a descendant of Miles Standish, excavated the site of his illustrious ancestor's house using remarkably careful techniques for the time.[3] Occasional other excavations were conducted on historical sites during the latter nineteenth century, and the pace quickened in the first half of the twentieth. But with relatively few exceptions, this work was motivated by a combination of antiquarian interest and a site's connection with some great American name, such as that of Thomas Jefferson or John Alden. Some projects were primarily exercises in the recovery of architectural data to aid in the restoration of historic sites, such as the program carried out in the 1930s and 1940s at Colonial Williamsburg.

It is only recently that historical archaeology has transcended this narrow perspective and become a useful contributor to the work of both historians and anthropologists. The fact that it serves two different disciplines—history and archaeology—has led to a dilemma of sorts: are historians and anthropologists equally qualified to conduct historical archaeological research? Genuine concern and a number of heated debates have sprung from this duality. Anthropologists often feel that historians have an overly particularistic approach to their data, whereas historians sometimes see a tendency toward overgeneralization and a disregard for the complexity of the past in the work of anthropological archaeologists. The fact remains, however, that historical archaeology in large part has been taught and carried out by anthropologists. Although there is nothing inherently wrong with this situation, historical archaeology needs both anthropological and historical perspectives to be fully effective. Anthropological archaeologists and historians often ask different questions. Neither are necessarily more "right" than the others; ideally they should be complementary and not opposed. From the outset historians and archaeologists work with different data bases: the historian with documents and the archaeologists with "material culture," "that sector of our physical environment that we modify through culturally determined behavior."[4] While the historian creates contexts of the past based on probate data, court records, censuses, diaries, and related written materials, the archaeologist's contexts are created from the study of excavated foundations, pottery fragments, faunal remains, smoking-pipe stems, and other such material realia. Since people in the past produced both documents and material objects, it is obvious that archaeology and history must be complementary. The real question is how best are we to combine the methods of historiography and archaeology to reach a better understanding of the past, not which of the two is more appropriate.

History's prime value to archaeology is a function of the richness of the documentary record. No amount of excavation can ever provide the kind of data used by historians to create a coherent, highly detailed construction of the past, and it should be against this construction that archaeologists project their findings. This does not mean that archaeology is simply a "handmaiden to history," as has been suggested by one eminent archaeologist.[5] Rather, while using the material record as a point of departure, archaeologists should seek explanations for their data in terms of the known history of the region and the time represented by their material. Such explanations can then be used to frame further questions to be asked of the archaeological data, and the answers to these questions again formulated with the historical record in mind.

Archaeology's prime value to history lies in its promise to take into account large numbers of people in the past who either were not included in the written record or, if they were, were included in a biased or minimal way. Slaves, indentured servants, poor tenant farmers, and modest freeholders formed the majority of the population in preindustrial America, but they were given less than full representation in the primary written sources. Even when they do appear, it is usually not their writing that we find but that of others, and one must take into account the biases of the recorder who was writing about them. As the folklorist Henry Glassie has so aptly stated: "A knowledge of Thomas Jefferson might be based on his writings and only supplemented by a study of Monticello, but for most people, such as the folks who were chopping farms out of the woods a few miles to the east while Jefferson was writing at his desk, the procedure must be reversed. Their own statements, though made in wood or mud rather than ink, must take precedence over someone else's possibly prejudiced, probably wrong, and certainly superficial comments about them."[6]

A second value of archaeology to history is a function of the commonplace quality of most material culture. As fundamental components of everyday life, things like houses, dishes, barrels, clothing, and food were so universal and taken for granted that there was little need to make written note of their existence, much less appearance. True, there are occasional building contracts, large numbers of detailed lists of household contents known as probate inventories, and other random mention of objects, but all fall short of the kind of detailed description required to make material culture useful in the construction of historical context. Archaeology has produced a rich corpus of closely dated evidence that, if used correctly, can provide insights not obtainable from the documentary sources.

The words "throughout the world" in our definition of historical archaeology take on particular importance when comparisons are made with prehistoric archaeology, the study of the immense period of time since people first appeared on the planet some three million years ago. Most prehistorians take as their study area a region of modest size, a river drainage perhaps, or a coastal strip not more than a hundred or so miles in length. Such regions conform to the extent of settlement by a particular prehistoric society, and it is such societies—archaeological "cultures"—that form the basic components of the prehistorian's construction of the human past. There are the occasional problems of prehistory that require a broader geographical scope, such as that of the peopling of America from Asia via the Bering Straits land bridge or studies of trading networks that extend over large areas. But by and large the prehistorian does not think on a scale that is truly global and has no need to operate on such a scale. Historical archaeology, on the other hand, must adopt a global perspective on its data, for when the first European sailing ships set out for distant parts of the world, a chain of events never before seen in human history was set into motion. Two worlds that had been separate from each other for millennia suddenly were brought into close contact, with spectacular and often catastrophic results. When one can hold in one hand a fragment of a tea cup made in Staffordshire, England, in 1810 and recovered from a well at the southern tip of the African continent and in the other hand its twin from the bottom of a trash pit in Massachusetts, this point is made with dramatic impact. It is thus not extraordinary that certain aspects of the archaeology of Flowerdew Hundred will require us to look at events taking place halfway around the world.

While archaeology will never replace history in the detailed accounting of past events, it can tell us things of value. As the historian Jack Larkin has told us: "Everyday life should not be exalted above other historical concerns. But it should not be ignored."[7] There are times when a better understanding of everyday life in the past can indeed lead to new insights concerning larger historical issues. And only archaeology can confront us with the actual physical remains of past communities—broken dishes, house foundations, pieces of hardware, and a thousand other kinds of objects that once had value to those who owned and used them. Flowerdew Hundred is not particularly special in this regard, for any Virginia plantation established in the early seventeenth century probably has an equally rich archaeological record. But over twenty years of archaeology at Flowerdew have provided the information to allow us to tell a story of considerable detail and interest which spans the two and a half centuries between Governor Yeardley's first settlement in 1619 and the end of the Civil War. That, then, is what this book is about.

Chapter Two

MORE THAN THREE CENTURIES of plowing have all but erased the physical remains of the houses, barns, and other outbuildings built and used by Flowerdew Hundred's first settlers. The method of raising these structures, known as earthfast, hole-set, or post-in-ground buildings, leaves the faintest of traces for the archaeologist to ponder. Buildings framed in this fashion rise from a number of large rectangular posts set deep into the ground, without benefit of sills or footings (fig. 3). Chimneys within the houses were also hole set, and the wooden frame of both house and chimney was then enclosed with woven sticks (wattle) and covered with clay (daub). The wattle and daub of the house frame was supported by smaller intermediate studs between the major framing members. Roofs were in most cases probably of thatch. Clapboard siding may have been used in some instances, attached to the frame with handwrought nails. Once such a building is razed or, for that matter, burns to the ground, the only evidence of its existence is the holes in which the posts were set. Plowing complicates the matter further, for the plow effectively removes any evidence in the top foot or so of the soil, making it necessary to dig below the plow zone, as this top layer is called, to undisturbed subsoil. It is there that the archaeologist can define the ephemeral faint shadow of the building that once stood on the site.

Finding and interpreting these stains in the ground is one of historical archaeology's most challenging and at times frustrating exercises. In con-

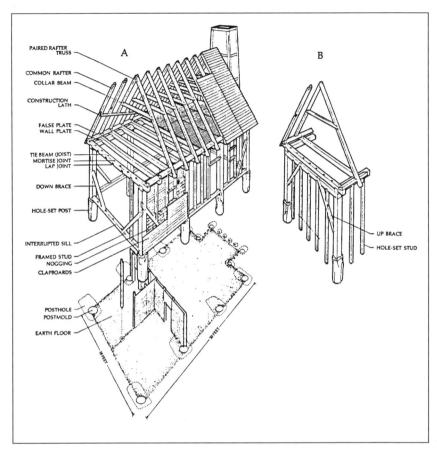

PAIRED RAFTER TRUSS
COMMON RAFTER
COLLAR BEAM
CONSTRUCTION LATH
FALSE PLATE
WALL PLATE
TIE BEAM (JOIST)
MORTISE JOINT
LAP JOINT
DOWN BRACE
HOLE-SET POST
INTERRUPTED SILL
FRAMED STUD
NOGGING
CLAPBOARDS
POSTHOLE
POSTMOLD
EARTH FLOOR
A
B
UP BRACE
HOLE-SET STUD

3. Typical framing system for an earthfast house and the posthole pattern that would be left in the ground. Frame with interrupted sills on left, hole-set studs on right. (Reprinted by permission of the publisher from Cary Carson et al., "Impermanent Architecture in the Southern American Colonies," *Winterthur Portfolio* 16, nos. 2–3 [© 1981 by The Henry Francis du Pont Winterthur Museum]: 143)

trast to the plow zone, which is usually a rich brown, the subsoil is a much lighter color. Any intrusion into the subsoil, be it by people digging pits or postholes or groundhogs burrowing into the soil, will leave a darker fill inside the hole. But *darker* is a comparative word, and often only slightly darker describes it better. This translates in practice as carefully cleaning the top of the subsoil, usually at a depth of a few inches below the base of the plow zone, removing the deepest scars left by the plow, examining large areas for the faintest of stains, and, once such stains are located, trying to discern the pattern that they form. A typical posthole appears as a stain, as large as three feet on a side, with a much darker stain within, the post mold marking the location of the post for which the hole was dug. The fainter mottling results from the hole having been dug several feet into the ground, bringing up clay which mixes with the soils from above. The mold is darker either because the post rotted in place or because it was removed, allowing darker fill to fall in from above. An area which has been cleared of the plow zone and allowed to dry in the sun will not show postholes; they appear only when wet, either when first exposed in the damp subsoil or when sprayed with water at some later time.

Newcomers to Chesapeake archaeology often think that people working out soil stain patterns are hallucinating, seeing things that are not really there. But they do exist, and by playing a dot-to-dot game, connecting the holes in various possible patterns, sooner or later we create a coherent picture and can say just what kind of a building stood on the site. A spectacular example of such careful and painstaking construction of post patterns is Frazer Neiman's excavation of Clifts Plantation in Westmoreland County, Virginia (fig. 4).[1] Neiman's work was nothing less than monumental, and involved the removal of more than ten thousand square feet of plow zone, to expose the remains of a fortified farmhouse and dependencies. One can easily imagine how impossible the job would have been unless a complete exposure was undertaken. An undergraduate student at Brown University once remarked, after working for days in the rocky soil around the Quaker meetinghouse in Newport, Rhode Island, "Archaeology could be so much fun if all the dirt didn't get in the way." Ten thousand square feet of plow zone translates into hundreds of tons of "dirt in the way," truly a formidable prospect. So what to do with it? Ask any three Chesapeake archaeologists how they deal with the plow zone, and you might well get three slightly different answers. But in general, there seem to be two schools of thought on the subject. The first holds that there is a large amount of useful information to be gained from some form of plow zone sampling, including screening it in its entirety, which to his great credit is how Frazer Neiman dealt with the plow zone at Clifts. Other archaeologists screen from 5 to 10 percent of the overburden by taking samples at regular intervals over the site. An even more radical position is taken by those at the opposite pole, who advocate wholesale removal with heavy equipment, a bulldozer or backhoe, and

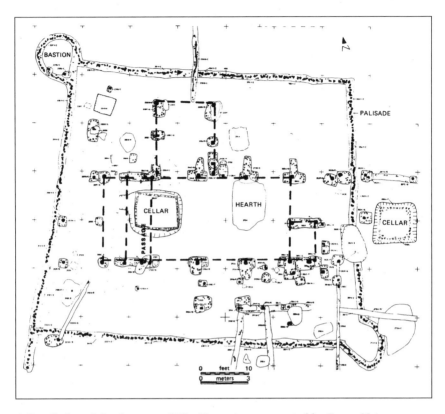

4. Fortified earthfast house at Clifts Plantation, excavated by Frazer Neiman. (Courtesy of the Robert E. Lee Memorial Association, Inc.)

no sampling at all. Justifications for both extremes are logical, and the whole matter eventually comes down to money and labor.

To be sure, there are traces of shallow features such as hearths, framing studs, or trash-filled depressions which, although totally obliterated by plowing, leave some material trace of their location. And while plowing can move objects from surface to subsoil and back over time, it seems not to displace them more than a foot or so laterally. So there is little question that such faint evidence exists. But the objects have lost all of their original context, while material from deep features in the subsoil have great contextual integrity. So it all comes down to whether one wants to expend the funds and effort to deal extensively with the plow zone or remove it rather roughshod to reach that portion of the site that has not been shredded by the plowshare. At Flowerdew the approach taken is a bit different yet, and has been shown to be quite effective. A careful surface collection is conducted on hands and knees on the site just after it has been plowed and disced, preferably after a good rain, which washes artifacts, bone, and shell clean of the clinging soil. Distributions are plotted within five or ten feet, based on a grid laid out over the site. Practice has shown that such a collection is roughly equivalent to a 5 percent screened sample, but the labor required is far less. Following such a collection, the site is scraped using heavy equipment to within an inch or so of subsoil. This remaining portion is carefully cleared with a trowel, and artifacts recovered supplement the surface collection, increasing the sample to roughly 10 percent. The troweling is also done to expose the subsoil and whatever soil stains it holds, so in fact two operations can be combined into one.

So, once the "dirt in the way" is dealt with, by whatever method, archaeologists are confronted with intriguing evidence of the earliest settlements in Virginia, Flowerdew Hundred among them. The early buildings at Flowerdew, along with comparable material from sites such as Kingsmill Plantation, Governor's Land, Clifts, and Martin's Hundred, have raised a series of important questions which take us into the broader sweep of events in early seventeenth-century Virginia. Recalling that pipe stem analysis suggests an early intensive settlement at Flowerdew, we must turn now to the sites that produced this evidence and the way they fit into what is known of the early history of the plantation.

Neither the exact origin of the name nor the precise date of initial settlement of Flowerdew Hundred is known. George Yeardley, who acquired the plantation in 1619, named it after his wife Temperance Flowerdew Yeardley, or at least after his wife's family. The word *hundred* probably designates a demographic-geographic division and in England was an area larger than a parish but smaller than a county. It has been suggested that in England it was used to designate an area of land capable of raising a hundred militiamen, but in Virginia clearly such was not the case. Several hundreds appeared along both sides of the James in the first years of settlement, and the name may have been given them as a propaganda ploy, making Virginia seem better than it truly was to those hearing or reading about it back in England. Although George Yeardley acquired the thousand acres that he named Flowerdew Hundred in 1619, it seems very likely that some settlement had begun there before that date, for his brother-in-law Stanley Flowerdew took a shipment of tobacco to England in the same year, probably grown on the same property. Flowerdew Hundred was certainly a fully established plantation by 1619, since it was represented in the first Virginia General Assembly in that year.

Unlike many of the very early Virginia plantations, founded as company establishments, Flowerdew Hundred was a "particular plantation," meaning that its owner had considerable latitude in managing its affairs and received at least a share of any profits realized. These particular plantations were a part of a more general reform of the colony's affairs intended to stimulate investment and settlement. In addition to Flowerdew Hundred, Yeardley also owned 2,200 acres across the James River at Weyanoke and maintained a residence in Jamestown. In fact, he seems not ever to have resided at Flowerdew, managing its affairs from a distance and visiting only occasionally. But Flowerdew Hundred grew under Yeardley's administration. It may have been heavily armed and fortified, for in the Indian uprising of 1622, only six people were killed there, and after that event Flowerdew was one of a small number of settlements that were not abandoned. Many of the colonists returned to Jamestown, and others consolidated in these few strongly defended settlements. In 1624 its sixty inhabitants were actively producing crops and livestock, including an annual tobacco crop valued at close to £10,000.

George Yeardley sold the plantation in 1624 to Abraham Peirsey who renamed it Peirsey's Hundred. Peirsey had come to Virginia as cape merchant (from the Portuguese *capo* meaning "head," thus head merchant) and soon became the second wealthiest man in the colony, just after Yeardley himself. A census taken in January 1625 (or 1624 by the Julian calendar) gives a clear picture of Peirsey's Hundred during his tenure, which was to be a brief one, since he died in 1628. This census, or muster as it was called, for an instant tears the veil of time from Flowerdew Hundred, making it suddenly come alive in sharp human detail.

THE MUSTER OF THE INHABITANTS
OF PEIRSEYS HUNDRED TAKEN THE 20TH
OF JANUARY 1624

Pierseys hundred
Ssamuell Sharpe arrived in the *Seaventure* 1609
Elizabeth his wife in the *Margrett and John* 1621
SERVANTS
Henry Carman aged 23 years in the *Duty* 1620
Peeces fixt, 2; Armours, 2; Swords, 2; Neat Cattell, 7; Swine, 6; Houses, 2.

THE MUSTER OF MR *GRIVELL POOLEY* MINISTER

GRIVELL POOLEY arived in the *James* 1622
SERVANTS
JOHN CHAMBERS aged 21 yeares in the *Bona Nova* 1622
CHARLES MAGNER aged 16 yeres in the *George* 1623
Corne, 8 barrells; Powder, 4 lb; Lead, 12 lb; Peeces fixt, 3: Armour, 1; Swords, 3; Neat Cattell, 1; Swine, 1.

THE MUSTER OF *HUMFREY KENT*
HUMFREY KENT arived in the *George* 16
JOANE his wife in the *Tyger* 1621
MARGRETT ARRUNDELL aged 9 yeares in the *Abigaile* 1621
SERVANTS CHRISTOPHER BEANE aged 40 yeares in the *Neptune* 1618
Corne, 6 barrells; Powder, 2 lb; Lead, 6 lb; Peeces fixt, 2; Swords, 2: Neat Cattell young & old, 3; Swine, 3.

THE MUSTER OF *THOMAS DOUGHTIE*
THOMAS DOUGHTIE arived in the *Marigold* 1619
ANN his wife in the *Marmaduke* 1621
Corne, 3 barrells; Fish, ½ hundred; Powder, 4 lb; Lead, 20 lb; Peece fixt, 1; Coat of Male, 1; Sword, 1; Swine, 1.

THE MUSTER OF *EDWARD AUBORN*
EDWARD AUBORN arived in the *Jonathan* 1620
Corne, 3 barrells; Pease, 1 bushell; Peece fixt, 1; Sword, 1.

THE MUSTER OF *WILLIAM BAKER*
WILLIAM BAKER arived in the *Jonathan* 1609
Corne, 3 barrells; Fish, 50; Powder, 1 lb; Lead, 12 lb; Peece fixt, 1; Coat of Male, 1; Sword, 1.

THE MUSTER OF *JOHN WOODSON*
JOHN WOODSON
SARAH his wife } in the *George* 1619
Corne, 4 bushells; Powder, 1 lb; Lead, 3 lob; Peece fixt, 1; Sword, 1.

THE MUSTER OF *EDWARD THRENORDEN*
EDWARD THRENORDEN arived in the *Diana* 1619
ELIZABETH his wife in the *George* 1619
Corne, 3 barrells; Fish, 1 hundred; Peece fixt, 1; Coat of Male, 1; Sword, 1; Swine, 1.

THE MUSTER OF *NICHOLAS BALY*
NICHOLAS BALY arrived in the *Jonathan* 1620
ANN his wife in the *Marmaduk* 1621
Corne, 4 barrells; Fish, 1 hundreth; Powder, 1 lb; Lead, 12 lb; Peece fixt, 1; Armour, 1; Sword, 1.

THE MUSTER OF *JOHN LIPPS*

JOHN LIPPS arived in the *London Marchannt* 1621
Corne, 10 bushells; Peece, 1; Armour, 1; Sword 1.

THE MUSTER OF MR *ABRAHAM PEIRSEYS* SERVANTS

THOMAS LEA aged 50 yeres

ANTHONY PAGITT 35
SALOMAN JACKMAN 30
JOHN DAVIES aged 45
CLEMENT ROPER 25
JOHN BATES aged 24
THOMAS ABBE 20 } arived in the *Southampton* 1623
THOMAS BROOKS 23
WILLIAM JONES 23
PEETER JONES 24
PIERCE WILLIAMS 23
ROBERT GRAVES 30

EDWARD HUBBERSTEAD 26
JOHN LATHROP 25
THOMAS CHAMBERS 24
WALTER JACKSON 24 } arived in the *Southampton* 1623
HENRY SANDERS 20
WILLIAM ALLEN 22
GEORG DAWSON 24

JOHN UPTON aged 26 on the *Bona nova* 1622
JOHN BAMFORD aged 23 yeares in the *James* 1622
WILLIAM GARRETT aged 22 in the *George* 1619
THOMAS SAWELL aged 26 in the *George* 1619
HENERY ROWINGE aged 25 yeares in the *Temperance* 1621
NATHANIELL THOMAS aged 23 yeres in the *Temperance* 1621
RICHARD BROADSHAW aged 20 yeares in the same Shipp
ROBERT OAKLEY aged 19 yeares in the *William & Thomas* 1618

Negro
Negro } 4 Men
Negro
Negro

ALLICE THOROWDEN
KATHERINE LOMAN } maid servants arived in the *Southampton* 1623
NEGRO WOMAN
NEGRO WOMAN and a young Child of hers.

PROVISIONS, ARMES ETC. of Mr PIERSEY at Peirseys hundred: Corne and Pease, 300 bushells; Fish 1300; Powder, 1½ barrell; Lead, 200 lb; Peeces of Ordnannce, 6; Peeces fixt, [*ink blot obliterates numeral*]; Murderes, 2; Armours, 15; Swords, 20; Dwelling houses, 10; Store houses, 3; Tobacco houses, 4; Wind Mill, 1; Boats, 2; Neat Cattell young & old, 25; Swine young & old, 19.

Mr SAMUELL ARGALLS CATTELL: Neat Cattell young and old, 8

DEAD at Peirseys hundred Anno Dni 1624

JOHN LINICKER
EDWARD CARLOWE
ROBERT HUSSYE
JACOB LARBEE
JOHN ENGLISH
CHRISTOPHER LEES Wife
ELIZABETH JONES[2]

Noteworthy in this accounting are a number of things. Dates of arrival, with the ships' names included, show how regularly people were arriving in Virginia. We tend to think of people of this time as relatively fixed in their place in the world, when in fact the seventeenth century was a time of great movement over vast distances. This is particularly striking when we realize that before the opening of the New World, most English people probably never traveled more than a few score miles from where they were born and where they would die. And this in turn makes us think of what it must have been like, so far from home, in a strange land trying to create anew at least some small part of that which was left behind. That they succeeded at all is no small wonder, and succeed they did, often with great skill and endurance. Six of the Negroes listed probably were some of the same ones who were brought to Virginia by a Dutch ship in 1619, for we know that both George Yeardley and Abraham Peirsey obtained fifteen of them. The seventh, the young child of an anonymous Negro woman, may well have been the first black child born on English soil in the New World and certainly was among the first few true African Americans.

Arms and armor, along with munitions, attest to the heavily defended state of the hundred at the time. "Peeces"—firearms—are sufficiently numerous that every male probably had at least one. Add to this the six pieces of ordnance and two murderers—small breech-loaded cannon that could fire anything from balls to nails—and the large quantities of livestock and staples, including great stores of fish, presumably dried, and a picture emerges of a highly successful establishment, well provisioned and strongly defended. One gets the impression that Peirsey managed the plantation more efficiently than did George Yeardley, and this makes one wonder if he was a more conspicuous presence there, even though he, too, maintained a residence at Jamestown. Both the archaeology and the records suggest that this was indeed the case. Certainly he paid continued attention to the defense of the plantation. In October 1626, less than two years after the muster, the colony passed a general law calling for all plantations to be palisaded. Peirsey was exempted from the law, as the records of the colony show.

13ᵗʰ day of *Octob* 1626,
A COURT at *James-Citty* the 13ᵗʰ day of *Octob* 1626,
present

Sʳ *George Yeardley* Knt Gouernoʳ &c, Capᵗ. *West,* Doctoʳ *Pott, Capᵗ. Smyth,* Capᵗ. *Mathewes,* Mʳ. *Persey,* Mʳ *Claybourne,* Capᵗ. *Tucker,* & Mʳ *fferrar. . . .*

3 *It is ordered,* according to an acte of a late generall Assembly yᵗ all dwelling houses through the Collony be palizadoed or paled about, defensible against yᵉ Indians to bee done & finish'd before yᵉ first day of *May* next. . . .

4 The Court at this time, vppon yᵉ demonstrance of Mʳ. *Abraham Persey,* yᵗ yᵉ aforesaid order would prove very heauye & burthensome vnto him at *Perseyes Hundred* is content, in reguard he hath many houses allredye paled & palizadoed in, & that all yᵉ whole necke is well railed in, & that he hath 10 or 12 pieces of ordinance well mounted & planted for yᵉ defense of yᵉ place, yᵗ hee doe pale or palizadoe in such other houses [as] are not yet secured from yᵉ Indians, as hee in his discretion shall thinke fitt.[3].

In March 1626, not long before he died, Abraham Peirsey executed his will. Referring to himself as "Abraham Peirsey of Peirsey's Hundred Esquire," he directed that "for my corporall bodie I bequeath that to the earth from whence it came to be decentlie buryed with out any pompe or vayne glorie in the garden plott where my new frame doth stand."[4] From this we know that he had built a house shortly before making out his will, but its location is not specified. Was it at Jamestown, or did he have it built at Flowerdew Hundred? If the latter, was Abraham Peirsey buried somewhere on the property? Archaeology has at least produced some intriguing clues on the matter.

Today the plantation that was once Flowerdew Hundred is known as Flowerdew Hundred Farm; the name has remained attached to the property continuously since Abraham Peirsey's daughter Elizabeth Stephens restored it at the time she acquired the property in 1636. It is shaped rather like a triangle, with the apex projecting northward into the James

River, which makes a wide bend around the property (fig. 5). This point of land is known as Windmill Point, so named after the fact that George Yeardley built what may have been English America's first windmill somewhere on his plantation in 1621; it is mentioned on the deed of conveyance to Peirsey in 1624. Flowerdew Hundred Farm was purchased by David A. Harrison III, a retired attorney, in 1968, and he resides there today. Harrison was aware that a windmill had once been on the property and, as anyone would, wondered just where it had stood. One area stood out as a good possibility. Near the tip of Windmill Point, farmers had been running into a mass of stone just beneath the soil as they plowed the fields, and the area around it was littered with fragments of glass, brick, and pottery. His curiosity piqued, Harrison engaged the services of archaeologists from the College of William and Mary in nearby Williamsburg to investigate further and to find just what lay beneath the surface. Serendipity led to the discovery of a remarkable complex of earthfast buildings, defensive ditches, and a variety of other structures, which have been under almost constant investigation to the present day. Although a complete picture is yet to be drawn, there is sufficient evidence at hand to permit some definitive statements about what is certainly the remains of the original Yeardley-Peirsey settlement.

Strung out along the shore of the James from northeast to southwest are a number of structures, some enclosed by palisades of one kind or another (fig. 6). At the northeastern end excavations revealed an enclosed compound consisting of at least two buildings and a well, entered by an opening some 25 feet wide through a palisade measuring 240 by 110 feet. Seven hundred feet to the southwest is a massive foundation of siltstone which once supported a house of considerable size. Just to the northeast of this building, a rectangular pit, 7 by 14 feet and 4½ feet deep, was uncovered and its fill removed. Between the enclosed compound and the stone foundation, there is good reason to suggest perhaps two or three other structures once stood, since surface collection and some preliminary clearing of subsoil give strong indications of their presence, though as yet this area has not been thoroughly investigated. Finally, at the far southwest end of the complex, a 40-by-60-foot palisaded area was uncovered, initially identified as an animal enclosure. From the start there was little question that this cluster of buildings was constructed quite early, and most evidence suggests that it took place before 1630. But more refined dates are called for, since Yeardley did not arrive at Flowerdew until 1619, and Peirsey was dead by 1628. Such accuracy is almost unheard of in

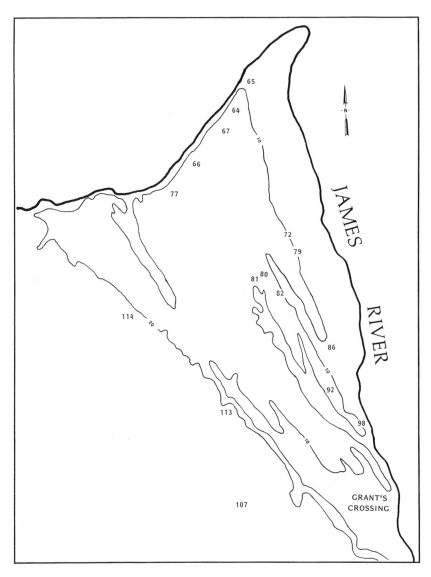

5. Map of Flowerdew Hundred, showing locations of sites mentioned in text

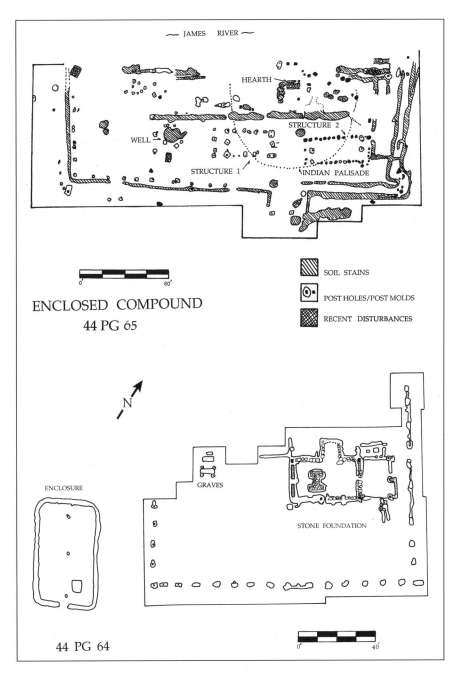

6. *Top:* enclosed compound; *bottom:* stone foundation, burials, and enclosure

historical archaeology unless some document can provide a precise date for a structure independent of the archaeological evidence. Yet, with a little imagination, some healthy conjecture, and a little help from our friends the artifacts, at least some suggestions can be put forth.

We begin with a German stoneware jar, of the type commonly referred to as a bellarmine or bartmann, or sometimes a graybeard. They are brown salt-glazed vessels, with a narrow neck, a loop handle, and a rim constructed to permit a cover to be tied on with strong string. Most are believed to have served at least initially as wine bottles. On the shoulder they exhibit a human face, with flowing beard, and those from the earlier years of their production frequently bear the seals of various European cities on their sides. They were manufactured in Cologne and later, after the 1620s, in Frechen. Halfway into the fill of the rectangular pit just to the northeast of the stone foundation, excavators found several pieces of a large bellarmine. When pieced together, they produced the top of the jar, looking rather like a funnel. But the bottom half was nowhere to be found. Such partially restorable vessels are commonly found in archaeological sites, and one assumes that the other pieces were scattered somewhere on the surface in the vicinity, not to be found without a bit of luck. However, in such cases one usually does not find exactly half of a vessel that pieces together, but rather odd assorted pieces, some of which fit together and others that do not. The pieces went into the lab along with all of the other material from the pit.

During the following winter a volunteer worker who was particularly taken with German stonewares decided to find the missing bottom half, although other staff members assured him that it was a useless effort. He made a systematic search of all of the brown stoneware fragments from every site on the property, but to no avail. Then, something wonderful happened. He noticed the bottom half of a large bellarmine that had been sitting in a case in the Flowerdew Hundred Museum, one that had been recovered from the enclosed compound to the northeast. When the two halves were brought together, they fit, producing an almost complete piece (fig. 7). The enclosed settlement is some 700 feet to the north of the pit from which the top half had come, prompting one person to suggest that the jar had exploded in midair, its pieces falling at a great distance from each other. But a more serious explanation provides the first clear connection between the enclosed compound, the pit, and the stone foundation adjacent. Numerous fragments of tile identical to those known to have been used to roof the house that once stood on the foundation were

7. Broken bellarmine jar: top section from pit structure, bottom portion
from enclosed compound, sites 64 and 65

found in the fill with the top half of the jar. One piece was that of a
capping tile, used to line the ridge of the roof, the only example of that
type recovered. The presence of these tile fragments in the fill of the pit
can be best explained as either the result of construction activities, being
scrap from roofing the house, or as the result of the house being razed,
when pieces of roofing tile might be expected to be strewn around the

area. However, the artifacts associated with both the bellarmine jar top and the tile fragments date mostly to the 1620s, while the house is known to have stood possibly into the 1640s, on the basis of pipe stem dates and other artifacts. These dates indicate that the presence of the tiles was not, therefore, the result of the house being torn down and allow us to suggest very strongly that the pit was filled as the house was built. Since the other part of the bellarmine was unearthed in the enclosed compound, we can reasonably assume that the construction of the house postdates that of the structures in the enclosure to the northeast. We are thus provided with a simple sequence of events, occurring between 1619 and 1630 plus or minus a year or two. The reason for the jar being found in two halves at some distance apart probably relates to the known scarcity of items in the early years of the colony. The half from the enclosed compound would serve perfectly as a bowl, and the other half could even have done in a pinch as a funnel. That such use of what to us would be badly damaged pieces of pottery took place even in more ample circumstances is demonstrated by Dutch genre paintings of the seventeenth century, which show badly broken plates and bowls stored on shelves along with undamaged ones.

At this point, there are but two facts on which to base any reconstruction of events in this early settlement, that there is an enclosed compound with at least two structures within it, and that nearby someone built a substantial house on firm stone footings. These facts are tied by evidence somewhat less substantial, but of use nonetheless. But a closer detailing of the archaeological evidence and a consideration of the relevant documents can move us further along in our understandings.

Semantics can be extremely important when dealing with archaeological matters, for often we are not sure of the function of the objects or buildings that we must deal with. It is for just this reason that prehistorians refrain from calling a triangular chipped-stone implement an arrowhead, but rather use the term *projectile point,* a seemingly equivocal word but one that nonetheless does not imply a use that is not known; it could just as well have been attached to a spear or dart. In like manner, the terms used to denote various early structures at Flowerdew have varied greatly according to the persons studying them. So before entering into a discussion of the various structures in the Yeardley-Peirsey complex, some sorting out of terms is in order. The set of two earthfast buildings within some kind of a fence or palisade has been variously referred to over the years as the "enclosed settlement," "fortified settlement," "fortified compound," "fort," and "enclosed compound." A fort it may or may not have been, and *settlement* implies somewhat more people than the evidence supports. *Compound* is, on the other hand, a nicely ambiguous word, and there is enclosure, as evidenced by the ditch surrounding it on all sides. Moreover, Ivor Noël Hume has used the term *compound* to describe a nearly identical complex at Martin's Hundred below Jamestown. For these reasons, the term *enclosed compound* is used here, since it is the least specific term, until further information is forthcoming. Likewise with the stone house foundation to the southwest. To be sure, the foundation is of stone, but what has happened over the years is that this foundation has more and more come to be referred to as the "stone house," the word *foundation* having slipped away. Of course, this designation is completely misleading, for it implies a house built entirely of stone, something never encountered in seventeenth-century Virginia. On the other hand, it certainly was a house. But in seventeenth-century usage, *house* does not necessarily denote a building in which people lived; there were cow houses, corn houses, and, of course, dwelling houses. Since we can be quite sure that the Flowerdew stone foundation is the remains of a dwelling house, the seventeenth-century term is quite appropriate. So *dwelling house* it is. And beyond the dwelling house lies the second enclosure, almost certainly not for animals, for in the seventeenth century one fenced animals out, not in. So *enclosure* will suffice for the moment. We are then left with three relatively value-free designations—enclosed compound, dwelling house, and enclosure—to describe the salient parts of the settlement associated with George Yeardley and Abraham Peirsey.

The dimensions of the enclosed compound are only accurate as to length. The north side is missing, the result of erosion by the James River, which appears to have removed as much as six to ten feet of shore line over the last three centuries and more. Much of this erosion has been recent, the result of strong waves left in the wake of oceangoing freighters traveling to Richmond and back. But in the 1620s the compound was certainly enclosed on all four sides, and relatively little seems to have been lost to wave action. Essentially rectangular in outline, with the slightest hint of a bastion in its southeast corner, it enclosed two buildings, a well, and a hearth, which might have been a part of a third structure (see fig. 6 top). Directly opposite the entrance gate can be seen a curved line of small postholes, evidence of an Indian palisade that had stood on the site before the English colonists arrived. These postholes serve as a powerful reminder that the history of Flowerdew Hundred does not begin in 1619. Evidence of occupation of Flowerdew Hundred's lands stretches back to at least 10,000 B.C. A Clovis point similar to those found on sites of early mammoth hunters in the West has been found near Windmill Point, and flakes of the same chert as that from which it was fashioned—a stone foreign to the area—have also been found in the plowed fields. These flakes may well mark the location of a deeply buried camp of early hunters. Between the time of the Clovis hunters and the seventeenth century, Flowerdew Hundred was the site of a continuous occupation by Indian peoples, from what is known as the Archaic period through to that time termed Woodland, from about A.D. 1200 through the late sixteenth century. By the time the first English settlers arrived, the local Indian population almost certainly had withdrawn from that part of the James River valley, leaving behind cleared fields that would be taken over by the Europeans. At Flowerdew Hundred this withdrawal might have been late indeed, for in 1982 archaeologists from the University of Georgia found brick fragments in Indian hearths in an apparently undisturbed context. These bits of brick could only have come from English settlements from the first few years of the colony's life, between 1607 and ca. 1612.

The two earthfast structures within the enclosed compound differ considerably from one another. Structure 1 is the remains of a typical earthfast building, with large posts set on approximately ten-foot centers, with a chimney at one end. This building was almost certainly a dwelling house. A shallow well was found and excavated ten feet to the west of the western end of the structure. This well produced a remarkable object, a perfectly preserved blackjack mug, made of leather sewn together and lined with pitch (fig. 8). It may be the only such object ever recovered archaeologically and is one of a small number that have survived to the present. The well water into which it was thrown or dropped formed a bacteria-free environment, accounting for its state of preservation. Structure 2, on the other hand, seems to have been a different type of building. It was constructed by setting puncheons—closely set smaller posts—between widely set larger posts, with a row of similar puncheons along each end. Lack of any clear evidence of chimneys or hearths has led archaeologists to conclude that structure 2 was probably one of the warehouses listed in the muster of 1625. On the east end soil stains suggest a loading ramp leading to the river shore, convenient for rolling hogsheads of tobacco to waiting ships.

These two buildings were enclosed by some kind of palisade or fence, although just what form this took is open to considerable question. A

8. Pitch-lined leather blackjack mug, from well in enclosed compound, site 65

continuous ditch surrounding the entire compound either held the bases of relatively light palings, tied together by horizontal rails, or more substantial thick sawn planks, tied in a similar fashion. The evidence as it stands allows little more in the way of clarification, but given the amount of military equipment found within the compound (fig. 9) and the size of the ditch—two feet or so across and slightly more than half a foot deep in the subsoil—it seems likely that the enclosure was intended to keep out more than errant cows or foraging hogs. If so, then it is a fortification of some kind, or at least some form of security against human entry, but whether from those outside the settlement or those within is yet another question. If structure 2 is a warehouse, the price of tobacco in the early

9. Military artifacts from enclosed compound, site 65. *Top,* breastplate; *left center,* "worm" for cleaning cannon barrels; *center,* cannonballs; *right center,* halberd blade (point down); *lower center,* musket rest. Six sizes of cannonballs were recovered from within the compound, matching the number known to have been present at the time of the 1625 muster but fewer than mentioned in the exemption given Peirsey from fortifying his settlement.

years may well have led to such security against people living at the site, with the palisade intended to safeguard a valuable commodity rather than the people who produced it. Or, given the perceived threat from both Indians and Spanish, it could have been intended as a defensive enclosure for the community as a whole.

Finally, it must not be assumed that the enclosed compound was all constructed at one time. If one plays dot-to-dot with certain features in the center of the compound, another possibility, highly conjectural, suggests itself. The hearth located near the river may or may not have been a part of a dwelling house. It is shown as one in a rendering of the compound in the National Geographic Society's book *Clues to America's Past*,[5] but that view is only conjectural. Still, the hearth is in an area of considerable erosion, and the long stain just south of it can be connected to certain posts on either side to produce at least a suggestion of a single house, possibly enclosed by some kind of pale. Should such have been the case, this single palisaded house may have predated the construction of the larger compound.

When we move southwest to the dwelling house, we are on a more solid footing, both literally and figuratively. As one of the earliest stone foundations in English America and the only one of its time in the Chesapeake, it is unique. No earthfast structure this, but rather a foundation that states with clarity just what kind of building once stood upon it. Forty-one feet long and twenty-four feet wide, with an eight-by-ten-foot porch on its northern side, the foundation is that of a hall-and-parlor house (see fig. 6 bottom). The floor plan is one of two rooms, on either side of a chimney, which is indicated by an H-shaped hearth offset to the west. This plan is found commonly in houses of more than one room built in America from the earliest years through the early nineteenth century and even later in some remote isolated areas. The larger of the two rooms, and the more public, entered directly from the outside, was a multipurpose space, where people cooked, ate, slept, and engaged in various crafts. The smaller room, the parlor, accessible only from the larger hall, was a more private space and usually served as a sleeping room for the head of the household and his spouse and a place in which to store and display more valuable items, such as clocks, plate, or special ceramics. Important guests may have been entertained there as well.

The remains of the foundation are massive semidressed pieces of siltstone, almost certainly imported from England, possibly Bristol. These stones show evidence of earlier use, and some have barnacles stuck to them. It may seem remarkable that a person would go to the trouble of shipping building stones from England to construct a house in Virginia. But if one wanted a substantial foundation of stone, no such material was available in the lower reaches of the James, and access to the nearest suitable material, west of the fall line and well into the piedmont, would have been difficult at the least. Once it is loaded on a ship, it makes little difference whether the stone travels ten miles or three thousand, for the real labor of loading and unloading it is the same in either case.

Certain details of the footings have generated considerable discussion regarding just what kind of a building once stood on the foundation. When the foundation was first excavated, it was thought that the building had been one of cruck construction. Cruck buildings, which were fast vanishing from the English landscape in the early years of the seventeenth century, are essentially like modern A-frame houses. Pairs of cruck blades, made by sawing a large tree lengthwise through the trunk and lowermost main branch, rather like butterflying a shrimp, serve as the main framing supports and carry secondary members, including the roof. This method of building contrasts with that known as box frame, in which major timbers are mortised together to form a box, upon which the roof is placed. Box framing is the tradition brought to America by the first English settlers, and as yet no cruck buildings have been discovered on the American side of the Atlantic. So it was with great excitement that the siltstone foundation at Flowerdew Hundred was perceived to be that of a cruck-built house. The evidence for such framing was a series of evenly spaced gaps, fourteen in number, which were thought to have received the bases of the cruck blades.

However, it must be remembered that the part of the foundation that we see today was well below the surface, and mortared brick in course atop parts of the foundation clearly indicates that the stones served as footing for brick walls that could well have risen several feet above the stone. When the house was standing, the stones were well out of sight, two or perhaps three feet below the surface. In the case of cruck building, the cruck blades do not extend into the ground, but rather are seated on stone pads. What, then, were the gaps for? Two interpretations can be put forth. The first is that they originally held brick piers, upon which the major framing timbers were placed. That they in some way relate to major house posts is clear from the way in which they are spaced, forming what are known as bays, those divisions of a frame house marked by major vertical framing posts. The alternative explanation, and the one more commonly subscribed to, is that the gaps held the bases of major posts, which in fact ordinarily did extend some distance below the surface. In either case, box framing is indicated, with continuous sills in the former and interrupted sills, tied to the posts, in the latter. The dwelling house that stood upon the foundation was probably two stories high, as indicated by the width of the footings, as well as the extension to the north, which may well have been a stair tower. The roof was clad with tiles, numbers of which were recovered in the course of excavation. At some later time further additions were made to the house, as indicated by soil stains to the north and east.

The dwelling house is situated within a yard outlined by massive square posts, set in holes from three to five feet square fronting the house and turning toward the river on either side. Within this yard were found three burials, dating to the time of occupancy. Two were the remains of males, each between forty and fifty years of age. The third, close beside one of the adult males, was that of a child approximately one year old. In a report prepared for the Flowerdew Hundred archaeologists, Douglas Ubelaker of the Smithsonian Institution identified the adults as European.[6] Such identification of infant bones is impossible, and even the attribution of origin of adults is not without problems. However, their proximity to the house, the method of interment—extended with heads to the west, typical of Christian burial the world over—and, most convincing, the fact that the two adults were lying in coffins held together with nails make it certain that they were Europeans. The coffins were of an unusual type. The lids were gabled, as indicated by a line of nails found down the midline of each skeleton, having dropped in a row when the wood of the coffin lid decomposed. Across the river below Jamestown, Ivor Noël Hume discovered five other burials with similar rows of nails down the midline. In *Martin's Hundred,* his engaging account of excavations at that plantation, Noël Hume gave a lively recounting of his extensive search for evidence of gable-lidded coffins of the type found at both Virginia hundreds.[7] More recently, in faraway Cape Province, South Africa, archaeologist Carmel Schrire found yet another gable-lidded coffin at the site of Oude Post, a late seventeenth-century Dutch fort. In this case, however, the nails were still in the position they had occupied above the body and were exposed in place, thanks to the excellent excavating skills of Cedric Poggenpoel.

The evidence then is clear: three people were buried in the yard of the dwelling house during the time it was occupied. One adult and the infant were interred in simple pits, but the other adult appears to have had some form of elaborate grave marker, indicated by four postholes, two at either end of the grave. Wooden grave rails were in common use in England at the time but usually were set parallel with the coffin, which would show as a single posthole at either end, on the midline. Perhaps these holes are the remains of wooden head and foot markers, set into the ground on posts. Whatever the case, it appears that someone went to rather elaborate measures to mark one of the graves.

Standing two stories high, with its red tile roof and massive brick chimney, the dwelling house must have stood out like a swan among dun-colored ducks. No other seventeenth-century building yet found at Flowerdew Hundred begins to approach it in substance, permanence, or elegance. That it may have been even grander than once thought is suggested by a set of three carved bricks recovered in 1989 from one of the postholes on the eastern side. These bricks are mortared together, plastered, and carved in an ornamental style that would be more at home at Hampton Court than on the raw Virginia frontier in the early seventeenth century. It will never be known certainly that this ornamental brickwork was a part of the house, since there is the slightest of chances that it came as ballast from England, but the probabilities are very high indeed.

As enigmatic as the foundation of the dwelling house is clear and explicit, the rectangular pit located just off the east side of the house, beyond the fence, admits of various interpretations. Its rectangular shape and even, flat floor make it seem more than simply a hole dug to obtain dirt or clay for some purpose or another. In dimensions and depth it is remarkably similar to the cellar house excavated at Martin's Hundred,[8] but unlike that feature, it lacks any evidence of posts within, although, as at Martin's Hundred, neither were there posts without, which one would expect were it a cellar for an earthfast house. It is possible that it represents an unfinished cellar house or, alternatively, one that was framed in such a way as to have left no traces of construction. From the fill a wide range of early seventeenth-century artifacts was recovered, including the bellarmine funnel. Most significant in terms of dating the fill were pieces of two jars, one which was fully restored, of a type known as West-of-England balustroid jars. Such jars are found infrequently on Cheseapeake sites, Martin's Hundred having produced several in a pre-1622 context. Others were recovered from the wreck of the *Sea Venture,* which went down off Bermuda in 1609, and they turn up at Fort Pentagoet, a French establishment in Castine, Maine, in a 1630s context. Other significant artifacts include a plate from a brigandine jacket, a kind of light armor made of small plates of steel riveted scalelike to fabric, large numbers of common pins, the lid from a brass bed-warming pan, and a number of tenterhooks. The tenterhooks may give a tiny clue to the purpose which the pit served. All had been subject to intense heat, which renders iron almost rustproof, and as such, they look almost new. Such hooks were used to stretch fabric on a frame, including canvas. With a little imagination, one might picture a very temporary pit dwelling, roofed with sailcloth stretched on a wooden frame, which may later have been burned, the hooks finding their way into the fill. Such a scenario would explain the lack of framing evidence within the pit. If it was a temporary pit dwelling, ceramic braziers or chafing dishes, common at the time, could have been used for whatever cooking was done. On the other hand, the pit may have served some very different function and still have been roofed with a canvas-covered frame. Whatever the pit was in its first life, it did provide us with the first evidence to relate the enclosed compound and dwelling house in time, in the form of the broken bellarmine jar.

A hundred feet farther southwest of the dwelling house a second enclosure stands as one of a pair of brackets framing the entire settlement (see fig. 6 bottom). At first thought to be some kind of pen for animals, this enclosure almost certainly served some other function, although just what that was is less than clear. Forty feet wide and sixty feet long, with rounded corners, the surviving trench extends an additional foot below subsoil in an area where the topsoil is particularly deep. This depth is matched by the depths of three postholes which are located at each end and at the midpoint of the long axis, in the center, and by a pit just inside the entrance and to the right. The best guess as to original depth of ditch, postholes, and pit would be of the order of three to four feet below the original surface of the ground. A palisade of such depth would be of considerable substance. Unlike the ditch surrounding the enclosed compound, the bottom of the ditch of this enclosure provided evidence of the way it was constructed. In a number of places could be seen the impressions of the bases of rather small triangular poles, presumably made by riving logs into quadrants. These appear to have been set quite close together and perhaps were gathered into bundles. Very similar impressions were found in the defensive ditch and bastions surrounding Clifts Plantation, and that enclosure has always been considered as defensive in nature. The function of the three large posts remains enigmatic, as does that of the pit inside the gate.

Artifacts were few in the ditch fill and pit; noteworthy are a completely restorable clay pipe of a very early type (fig. 10), lead net sinkers, a fishhook, and small buckshot. Splashes of melted copper and bits of lead ore hint at some kind of smelting activities. That such activities were taking place at Flowerdew is further suggested by pieces of crucibles discovered

10. Smoking artifacts: brass pipe tamper from enclosed compound and complete early pipe from defensive palisade trench, enclosure

in the enclosed compound. While it would be quite useful to be able to say more about this enclosure, we are not at a total loss. It can be said with considerable assurance that it is not an animal enclosure, both because animals were not penned at that time and on the basis of evidence that other activities took place within. There is a decidedly male flavor to the few things recovered, although gender attributions to seventeenth-century objects and structures are fraught with difficulty. Women smoked pipes, tilled the soil, and engaged in a whole range of activities that one might uncritically see as traditionally male. In fact, female-related artifacts, those associated with food processing and preparation, are probably more unequivocally gender related than such things as pipes, needles and thimbles, firearms, or agricultural tools. When one considers the very unbalanced sex ratio in early Virginia, the matter becomes even more complicated, for with so few women in the early years, men probably took on tasks that later would, for better or worse, fall to women.

If we reduce this settlement along the James to its most fundamental functions in almost a grammatical fashion, like parsing a sentence into component parts regardless of the specific words involved, we arrive at a basic structure of few parts. The functions served by this settlement are but three in number: defense, commercial enterprise, and domestic life.

All three are indicated by the various structures in the settlement. The compound was certainly at least partially a commercial center, if as most people agree, structure 2 is a warehouse with docks nearby. It has elements of defense as well, although against precisely whom is not clear. The dwelling house is but one of a number located in a line between compound and enclosure, if surface indications and limited excavation are at all correct in their implications, as they certainly seem to be.

Thus the domestic section of the community lies between a pair of palisaded enclosures, both having some defensive function. It is not essential that these functions be spelled out in detail, for in general structure and layout the settlement is by no means unique, but rather matches in many ways communities both in early English America and Northern Ireland. In *Martin's Hundred* Noël Hume called attention to the parallels between frontier communities in Ulster and the Martin's Hundred settlement plan, all sharing in similar solutions to the three linked needs of business, defense, and daily living. The Ulster settlements were established under James I in 1609 and were underwritten by various London merchant guilds. As such, they were first and foremost commercial enterprises. But there was in Ulster, as in America, a need for defense, there against the "wild Irish" beyond the pale. Each was to include at least a "bawn," a fortified enclosure within which the leader's home was located, but which would also serve as a place of refuge for the settlement as a whole should need arise. Dwelling houses were arranged along a street leading away from the bawn's entrance. Flowerdew Hundred and Martin's Hundred are laid out in like manner, and other early seventeenth-century settlements in English colonial America follow suit. In his important work *Architecture and Town Planning in Colonial Connecticut*, Anthony Garvan was the first to point out the relationship between town layouts in the first years of settlement in the American colonies and those of Ulster.[9] Plymouth, Boston, and Sagadahoc, Maine, all conform to the model. It would seem that the various companies underwriting settlement in both Northern Ireland and North America had a clear plan of what a proper community should look like and saw to it that they took a generally similar form. As Garvan noted, this would be no surprise since half of the merchant adventurers who underwrote these settlements had shares in both Irish and American enterprises.

These relationships and convergences assume critical importance when brought back to ground along the James at Flowerdew. If we had only the Flowerdew evidence on which to base our impressions of what the early

Yeardley-Peirsey community looked like, filling in the many gaps that remain would be quite difficult. But when the parts that are known fit so
comfortably in a more general pattern, we can postulate the existence of
the other dwelling houses even in the absence of archaeological data that
indicate their presence. And if only the enclosed compound were known,
with no comparative examples, arguments would rage endlessly over
whether it was a fort, a commercial compound, an enclosed settlement,
or whatever else one cared to call it.

Not all of the people living at Flowerdew Hundred during the first half
of the seventeenth century resided in the Yeardley-Peirsey settlement. At
least five other locations had some kind of buildings standing on them,
mostly dwelling houses, judging from artifacts collected from the sites.
All appear to have been of earthfast construction, for plowing has failed
to turn up even the slightest trace of more substantial foundations. Two
of the seven sites have been excavated in part; the remaining five are
known only from material collected from the surface. These are the sites
which produced the pipe stem collections that indicate an initial buildup
of settlement in the years before 1660. In all likelihood each of these
dwellings had a name of some kind, in the manner of the time, perhaps
something as fancy as Auborn's Pride or as plain and straightforward as
Mr. Baly, his house. But the names are lost to time, and today the same
locations have been given sterile designations such as site 44PG68 or site
44PG72 (see fig. 5). The number 44 refers to the state of Virginia, forty-
fourth in the alphabet before the admission of Alaska and Hawaii (which
are numbers 49 and 50). Prince George County, in which Flowerdew Hundred is located, becomes PG. The final number or numbers refers to the
order in which the site was surveyed and registered with a central state
agency. As such, there can only be one site 44PG67, as well as only one
site 39ST2, in South Dakota, or one site 4SBa32, in California.

The dwelling house and enclosed compound in the Yeardley-Peirsey
settlement were given two numbers, 64 and 65, although now it is apparent that they are each a part of a larger, single complex. The five other
sites are fairly regularly spaced along a line running north-south across
the bottomlands of the farm, almost as if they were connected by a road
or pathway, as they may well have been. Site 68 is just inland from the
riverside settlement. Known only from surface materials, it appears to
have been occupied during the second quarter of the seventeenth century
and perhaps a decade into the third. Ceramics, which along with pipe
stems provide the best indicator of the date of a site, bear this out. The

pottery collected from the site includes quantities of early Frechen and possibly Cologne brown stoneware; Wanfried slipware, a decorated German pottery common in the first quarter of the century; and a tin-glazed pottery commonly known as delftware. The site also produced an Elizabethan sixpence and a pistol barrel. Farther south, site 72 was excavated in 1973 and 1974, revealing an earthfast building forty-three feet long and sixteen feet wide. Lacking evidence of any kind of chimney or hearth and providing a very small number of artifacts, it is thought to be the remains of either a barn or a warehouse. This site has the distinction of having the earliest pipe stem date of all the Flowerdew sites, with 32 percent of the bores measuring $\frac{9}{64}$ inch, 62 percent $\frac{8}{64}$ inch, and only 6 percent $\frac{7}{64}$ inch. As such, it must be one of the structures referred to in the muster of 1625.

Site 79, to the south of site 72, seems quite early on the basis of a barely adequate sample of pipe stems, and the ceramic evidence accords with this supposition. A single roof tile fragment offers the slimmest of clues to the presence of a house on the site. Site 79 also has produced the only artifact to date that could have a direct connection with George Yeardley (fig. 11). It is a brass medallion, bearing the bust of a gentleman in armor and ruff, with the date 1615. Circling his head is an inscription in Latin: "MAVRITIVS AVR. PRINC. COM. NASS. ET. MV. MAR. VE. FL. EQ. OR. PERISCLIDIS"—Maurice, Prince of Orange, Count of Nassau and Merur, Marquess of Vere and Flushing, Knight of the Order of the Garter. The medal was struck to commemorate Maurice's induction into the Order of the Garter in 1612. He was a staunch ally of England in the wars in the Netherlands, a Protestant hero in the war against Spain and Catholicism. In 1601 sixteen-year-old George Yeardley was among Maurice's troops. Since such medals were freely distributed on the occasion of election to the order, Yeardley could well have come by one and carried it as a talisman to Virginia. Three other such medallions are known, one of silver and one of brass in the British Museum and a brass one from an Indian grave in Rhode Island.

Site 82 is next, to the south of site 79. Excavations here have produced evidence of an earthfast dwelling house and an unusual pit structure which appears to have served as an oven, possibly for baking bread. The pit is roughly circular, eight feet in diameter and three and a half feet below subsoil. In one wall is a small oven of highly fired clay, with an opening one and a half feet in diameter and a depth of three feet. The sides and floor of the oven are fired brick hard and are red in color, and

11. Maurice of Orange medallion

the arched roof is of a similar hue but not quite as hard. Layered deposits of charcoal and ash outside the mouth of the oven indicate repeated use. It certainly could have served well as an oven for baking bread; similar ovens are in use even today in rural areas the world over, though they are not usually set in the wall of a subterranean structure. A fire is built inside, the oven is brought to the proper degree of heat, and the ashes are raked out. The loaves are then placed within the oven, the opening is closed, and the bread bakes from the residual heat. Such a function for this oven seems most likely, although other possible uses such as drying corn should not be ruled out. The records lend additional support to this interpretation, for in 1621 the Virginia Company issued instructions to the governor and Council of State in Virginia that corn mills and public bakehouses be constructed in every borough. That a windmill stood at Flowerdew Hundred from at least 1624 is known from both the deed of conveyance and the 1625 muster. The existence of a bakehouse as well is suggested solely by the archaeological record. Site 86, near the shore at the southern end of the farm, completes the set. It has not been excavated, but both pipe stem dates and ceramics suggest a quite early beginning date, somewhere in the 1620s. Fragments of roof tile provide evidence for a dwelling house on the site.

The pieces of the puzzle have now been laid out, and it is time not only to fit them together but to attempt to see how they relate both to events and people at Flowerdew Hundred before 1660 and to broader questions about the nature of Chesapeake society during the same period. We will work outward, first providing Flowerdew with some coherence and then fitting the plantation into the society of which it was an integral part.

From an archaeologist's point of view, the brief time segments of the Yeardley-Peirsey years are practically impossible to separate, given the imprecision of the various dating techniques available. On the other hand, with the help of documents and certain relative time relationships—i.e., A was built before B, but we don't know exactly when A was constructed—a reasonable scenario can be offered. And it is often better to risk being wrong than to retreat into timid equivocation. The first and most vexing question to be posed is that of Flowerdew Hundred's physical appearance in the brief time between 1619 and 1624 when it was sold to Abraham Peirsey. The 1625 muster gives us a lucid and detailed description of the plantation, which fits the archaeological evidence with remarkable precision. The muster lists some seventeen structures, ten of which are dwelling houses. We have archaeological evidence for a minimum of seven dwelling houses and two other buildings, the structure at site 72, probably a barn or warehouse, and the warehouse within the enclosed compound. There were certainly other dwelling houses between the enclosed compound and the dwelling house in the riverside settlement, perhaps three. This would bring the total number of dwelling houses in line with the muster, and if the odd barn or storehouse stood adjacent to one of the dwelling houses south of the settlement, the fit would be almost perfect. By 1626 Peirsey was able to claim that he had "many houses allreadye paled & palizadoed in" and that he would enclose the others, probably those farther to the south, when he saw fit. In short, the archaeological evidence and the muster are in agreement for the years 1625–26. If we did not know that the plantation had been previously owned by Yeardley, the entire set of sites could be ascribed to the hand of Abraham Peirsey. But that leaves five critical years unaccounted for, and we must ask to what degree did George Yeardley develop the plantation before selling it to Peirsey. Another company-decreed census, "The Lists of the Livinge and Dead in Virginia," taken eleven months before the muster of 1625, in February 1624, places sixty people at Flowerdew Hundred, including eleven Africans.[10] But we must remember that this list was made less than two years after the uprising of 1622, and Flowerdew Hundred's population had been considerably increased by people coming in from other plantations, those that had borne the brunt of the attack far more severely than had Flowerdew. Which brings us inevitably to the events of that March morning in 1622.

Relations with the local Powhatan Indians during the first twelve years of the colony were turbulent and marked by intense hostility on both sides. Colonists would rather steal corn from the Indians than grow their own, and the expected reprisals took place. This changed in 1619, when the company established a wide range of policies intended to reform what had gone before and set the colony off on a proper course, perhaps for the first time. These reforms included the establishment of particular plantations, including Flowerdew Hundred, and a rather idealistic effort to accommodate the Indians into the grand scheme of things. Funds were set aside to establish an Indian college in Henrico, near modern Richmond, and whole Indian families were taken into English communities to learn the ways of proper Christian civilization. Opechancanough, who in that year succeeded Powhatan as supreme chief, seemed to go along with the plan; and the settlers, at last relieved of their earlier fears of attack, spread out along the James, creating numbers of new settlements.

They were rudely surprised on the morning of March 22, 1622, when the Indians among them fell upon them with whatever makeshift weapon might be at hand—a hoe, hammer, spade, or ax. By noon, 347 settlers were dead, mostly killed with the very implements that they had been using to maintain their lives. Such an attack, from within rather than from without, would have succeeded even if a settlement had been surrounded by batteries of cannon, which calls into question just why Flowerdew Hundred had such a small loss of life, only six individuals. The list "of the names of all those that were massacred by the treachery of the Sauages in Virginia, the 22nd March last" shows little pattern, although the heaviest losses seem to have occurred north of the river and to the east.[11] Thus Martin's Hundred, downriver from Jamestown, lost the most people, seventy-eight in number, and at Weyanoke, Yeardley's property just opposite Flowerdew Hundred, twenty people perished. Against this, one wonders whether Flowerdew's supposed heavy defenses in 1622 had anything to do with the small loss of life there. It is not clear whether the six pieces of ordnance listed in the 1625 muster were in place as early as 1622; Peirsey had doubled that number in only a year when his court testimonial was recorded.

If we assume for a moment that Flowerdew Hundred's population was roughly doubled after the 1622 uprising, not unlikely considering that only seven settlements absorbed all of the people coming in from other places, we get a figure of somewhere between twenty-five and thirty-five people there before March 1622, not an unreasonable estimate. The fifty-seven souls living at Flowerdew in 1625 were sheltered in ten dwelling houses, which works out to slightly more than five persons per house. By seventeenth-century standards of shelter for servants and tenants, this is quite commodious. Servants usually lived in tight quarters in attics, or sometimes in a special portion of the house separated from the main quarters by a cross passage. Such a house was found by Frazer Neiman at Clifts Plantation, and numerous other examples are known in the area, dated to the same period. If we apply the figure of five or six people per dwelling house to a projected population of thirty for Flowerdew on the eve of the uprising, we can suggest that a maximum of six houses or as few as four would have been more than adequate, especially considering that except for the minister, Grivell Pooley, those at Flowerdew at that time were no doubt all servants, and probably included in their numbers some of the Africans who appear in "The Lists of the Livinge and Dead."

Further support for this argument comes from the people themselves. When the lists of 1624 and 1625 are compared, we find that only fourteen of the fifty named people living at Flowerdew under Yeardley's administration were still there after Peirsey bought the plantation. Curiously, four of them are listed as Peirsey's servants. Of the remaining thirty-six named individuals, six are listed in the muster as residing at Yeardley's holdings at Hog Island, and another ten with Yeardley at Jamestown. Six others appear in muster lists from all over the colony, not more than one in any single location. They scattered to Charles City, West and Shirley Plantation, James Island, the Eastern Shore, Elizabeth City, and Newport News. Two others appear in early land transactions, including Lieutenant Gilbert Pepper, who patented property on the Warwick River in 1627. George Yeardley had paid for Pepper's transportation to Virginia. Twelve names do not continue to appear in any of the contemporary records, and those people may well have returned to England.

Thus "The Lists of the Livinge and Dead" do not reflect the true size of Flowerdew Hundred as it was before the uprising, since the population was swollen by refugees from other parts of the colony. And even sixty people could have been sheltered in four or five houses under the less than ideal circumstances that must have prevailed in the days and weeks that followed. By all appearances, then, Yeardley's Flowerdew Hundred was a significantly smaller settlement than that established by Abraham Peirsey by 1625. The riverside settlement could well have accommodated the thirty-odd people living there before 1622. In view of the very early dates that we can assign to at least a part of that settlement, this is not at all at odds with the archaeological evidence. It is impossible to resist the temptation to add up these numbers in the following manner. If we assume that the people who were at Hog Island and James City, all Yeardley's, had been long-term residents at Flowerdew Hundred, as well as those who appear at Flowerdew in both "The Lists of the Livinge and Dead" and the 1625 muster, we have a total of thirty. Not included in this accounting are the ten "Negros" who appear in the 1624 list; their number would raise the total to a maximum of forty, a little more than two-thirds the size of the population at Flowerdew Hundred in 1625. The remaining seventeen people are the most likely to have sought shelter at Flowerdew in 1622 and left sometime during the period between the taking of the two lists.

It is virtually certain that Peirsey built the two-story dwelling house to the west of the enclosed compound. The broken stoneware jug from the filled pit near the house allows us to say that the house was built after the enclosed compound with its two or three structures, but how much later is still a question. Other archaeological and documentary evidence would suggest a construction date sometime between 1625 and 1626. In the 1625 muster Peirsey is still listed as residing at Jamestown, with two maidservants and two manservants. By March 1626, when he drew up his will, he explicitly referred to himself as "Abraham Peirsey of Peirsey's Hundred," which would strongly suggest that he was residing there at the time. Add to this his references to his "new frame" and "garden plott," and the identity of the person buried in the elaborately marked grave just west of the house is no longer a complete mystery. Sex and age match the facts—Peirsey was fifty-one when he died in 1628. The fit between archaeology and the documents is comfortable in this case, although as so often is the case in matters of this sort, we will never be absolutely certain. However, the staff at Flowerdew Hundred felt certain enough to return the bones to the ground in the summer of 1989 on the assumption that they were those of Abraham Peirsey. On a twilight summer evening, the remains were committed to the ground in a solemn service in keeping with his period. Passages were read from the Book of Common Prayer by both a lay Episcopal minister (an Italian-American from New York) and a Presbyterian elder from South Africa. A bell was rung once for each year of the person's life. The bell tolled fifty-one times.

Not included in our accounting of the archaeological evidence from the Peirsey years is what may have been his most ambitious construction project, the railing in of the "whole necke." No line of great posts has been uncovered, and indeed the exact boundary of the "whole necke" is far from clear. Yet a few tantalizing scraps of evidence suggest that small portions of such a great palisade have already been located but not identified as such. The line of massive posts that encloses Peirsey's house, set in holes as large as any found on Chesapeake sites, seems rather formidable to enclose but a single dwelling house. True, it makes a return toward the river on the west, but there is as yet no evidence that it does not also continue beyond. The enclosure lies just past the last hole in the line, and as yet no excavations have been conducted immediately beyond that feature. However, excavations have been carried out some distance beyond the enclosure, and these have turned up what could be the continuation of the line. At site 77, a quarter of a mile southwest of the Peirsey house, archaeologists cleared an extensive area in an attempt to locate postholes that went with a rather large tile-floored cellar of the second half of the seventeenth century. While this effort was somewhat less than successful, the area exposed revealed two sets of stains that were not aligned with the cellar in compass orientation. One set was a row of late eighteenth-century burials which were not excavated, although a test in one of them, to confirm this identification, produced a whole clay pipe from the fill. The other set of stains were those of posts identical to those fronting the Peirsey house, set on the same ten-foot centers. These posts turned a right angle toward the river in the area exposed. Two hundred feet beyond the enclosure, a test excavation was made in the hopes of finding nothing, an unusual archaeological exercise. The reason for this test was simple. This area near the settlement had been suggested as a possible site of a reconstruction of the Peirsey house, and it was necessary to make certain that no archaeological remains would be damaged. Remains there were, however, in the form of four large posts, set in a rectangle on ten-foot centers, identical to those at both the Peirsey house and site 77. When a transit was set up and the alignment between the three sets of posts was checked, they lined up perfectly. If all three sets are parts of a single long line of posts, it follows that Peirsey's rail has been found.

In many ways such a palisade makes sense, for the other, eastern side of the neck is a cypress swamp, almost impenetrable, a natural barrier which with a palisade along the western side of the neck would serve well to protect the settlement. Although fear of attack from the water by either the Spanish or pirates was always present, it is clear that the defensive precautions specified in the 1626 court order were generated by a fear of land attack by the Indians, with the memory of March 1622 still fresh in people's minds. To date, it has not been possible to coordinate labor resources and planting schedules in order to expose the entire line, if such exists. Large-scale excavation would be essential, and this can only be done during winter months, but archaeology is a summer pastime by and large. In time, however, we may unearth an entire line of palisade postholes.

What, then, has the archaeology told us that we would not have known if the site surveys, excavations, and artifact studies had not been done? In truth, not all that much, but this is really not what is at issue. What archaeology has done is confront us with a body of material evidence which in turn has led to asking different kinds of questions of the written sources. And the material world of men like Yeardley and Peirsey possesses a powerful emotional content not to be dismissed. Finally, at a more general level, archaeology has made an important contribution.

The attempts to sort out the Flowerdew archaeology into Yeardley and Peirsey components has raised questions about the men themselves. Looking at the archaeological evidence for those early years, we are struck by the impact Peirsey made on the landscape, effectively masking that of the earlier Yeardley years. The documents bear this out as well, but there is something very reassuring about tying these records into the real tangible world. So it is that taken together, archaeology and the written word are far more persuasive than either would be on its own. With this assurance, we can engage in a little psychohistory and speculate on what motivated both early owners of Flowerdew. Neither man started out wealthy, but by the time Yeardley was knighted in 1617, an action that he apparently actively promoted, he was quite well off, and he returned to Virginia in the most comfortable of circumstances. There is something of the vain-glorious about the man, and from a distance of three and a half centuries, he still seems rather full of himself. While in London after obtaining his knighthood, he was observed to "flaunt" his new title "up and downe the streets in extraordinarie braverie, with fowreteen or fifteen fayre liveries after him."[12] He is mentioned in "The Lists of the Livinge and Dead" as George Yeardley, Knight. His business interests were diverse and wide-spread, and Flowerdew Hundred seems not to have been at the top of any of his enterprises. He was assigned 3,000 acres and 100 tenants upon assuming the governorship in 1618, and of these tenants possibly as few as 30 resided at Flowerdew. Characterized by his contemporaries as a "right worthie Statesman for his owne profit" and a "man wholy adicted to his private,"[13] he was always surrounded by controversy and accusations of various wrongdoings, from teaching the Indians how to use firearms to refusing to return all of his tenants upon relinquishing the governorship in 1621. Since he never resided at Flowerdew Hundred and was involved in such a variety of dealings of a proper and not so proper nature, it is almost certain that the plantation never received his undivided attention.

Abraham Peirsey is a very different story. His life in Virginia can be bracketed by two contemporary quotes; a "verie poor man" upon arriving in the colony, after only twelve years he "left the best Estate . . . ever yett knowen in Virginia."[14] To be sure, he was well connected, being a distant relative by marriage of Queen Elizabeth, and he was certainly not above the occasional shady business deal. But the fact remains that he rose to his position as second wealthiest man in the colony without benefit of a knighthood or commission of governorship and all the accompanying perquisites. A seventeenth-century self-made man, Peirsey probably knew the value of hard work and perseverance, and unlike Yeardley, his commitment to his hundred on the James seems to have been intense. Not only did he rename it after himself, but he saw to its defenses in a thorough way and, in all likelihood, built himself a house and moved to Peirsey's Hundred two years before he died. He seems to have left the estate in reasonable shape, for even though court litigation kept it from passing to his daughter Elizabeth Stephens until 1636, the occupants of the river settlement and the seven outlying sites stayed past mid-century, apparently engaging in profitable tobacco production, and still others remained to diversify their economic interests in new directions.

Earthfast building and economics are linked closely in a complex fashion. In an important article, "Impermanent Architecture in the Southern American Colonies," Cary Carson and his colleagues, both archaeologists and historians, have presented a strong argument linking earthfast building to tobacco monoculture.[15] They took issue with Edmund Morgan's argument that such impermanent building was a sign of a "get rich and get out" mentality during the years of the tobacco boom, pointing out that such construction continued well into the eighteenth century in parts of the Chesapeake. Their thesis, in turn, has been questioned by Robert Saint George in a paper entitled "Maintenance Relations and the Erotics of Property." Saint George argued that impermanent, maintenance-intensive buildings constitute an important part of the social contract that binds a community together. All three arguments have merit and can easily be combined into a single covering explanation.

Carson and his coauthors explained that earthfast building has roots deep in time in England but by the seventeenth century had almost vanished, giving way to more substantial "fairly framed houses." But the seeds of the knowledge to build such structures were carried across the Atlantic and, when planted there, flourished with amazing vigor. Recent reassessments of surviving seventeenth-century buildings in Virginia and Maryland leave us with but a scant half dozen in the entire Chesapeake, and these mostly date to the very end of the century. Only one is a hole-set structure. By contrast, New England can boast seventy-one buildings surviving from the same period, with ten predating 1660. Here archaeology has produced the critical evidence to account in part for the difference. The basic reason is quite simple; in the Chesapeake, earthfast buildings were almost universal; they were employed even by the wealthiest people in the region. There are a few exceptions, among them St. John's at Saint Mary's City, Maryland; Bacon's Castle in Virginia; and, of course, Abraham Peirsey's "new frame" at Flowerdew. But for the most part, the entire architectural tradition of the seventeenth-century Chesapeake has vanished from sight.

True, this type of building did not require archaeology to prove its existence, for there are early records that describe such buildings in some detail, as Edmund Morgan pointed out:

> Visitors to Virginia rightly judged the intentions of the settlers from the way they were content to live. "Their houses standes scattered one from another, and are onlie made of wood, few or none of them beeing framed houses, but punches [posts] sett into the Ground And covered with Boardes so as a firebrand is sufficient to consume them all." In fact, it did not even take a firebrand. Virginia "houses" could be kept standing only with difficulty. At Charles City, where the settlers had considered themselves fortunate to be released earlier than others from the company's service, they went on building "such houses as before and in them lived with continual repairs, and buildinge new where the old failed." These was no point in putting up more than a temporary shelter if you did not intend to stay; and as late as 1626 the governing council admitted that what people looked for in Virginia was only "a present Cropp, and their hastie retourne."[16]

But these quotes are from the colony's early decades, and what archaeology has shown is that earthfast construction far outlived the tobacco boom, which ended abruptly in the 1660s when Charles II imposed revenue taxes. Carson and his colleagues rightly questioned why the technique of earthfast building lasted so long in the South, while not appearing to any significant degree elsewhere. Their explanation relates to the way in which one manages limited resources in the most efficient fashion. Tobacco is a highly labor-intensive crop, and labor, either through purchased indentures or purchased Africans, was expensive. A costly substantial house would therefore have a lower priority than the labor that would produce a paying crop of tobacco. Tobacco remained the main if not exclusive crop in the Chesapeake through the seventeenth century and in some areas, including the Southside where Flowerdew Hundred is located, well into the eighteenth. They divided the Chesapeake into twelve subregions and, in each, showed that when tobacco ceased to be the major or sole crop cultivated, construction changed from earthfast to more substantially framed buildings on firm foundations of brick. Even houses made entirely from brick appear at this critical juncture. The change took place earliest in lower Norfolk County, where the shift from tobacco to mixed crops occurred earliest in the region, around 1680. As they suggested, it must be more than coincidental that the earliest small brick houses are to be found in the same area. And so it goes, in subregion after subregion; although the date of changing from earthfast building to more substantial construction varies from place to place, it coincides every time with the change from one kind of agricultural production to another. Wheat especially needs but a small portion of the labor required to produce tobacco. The planters, small and large alike, were freed from such an intense commitment of labor to tobacco production. As a result, resources could be redirected, and apparently the building of a better house took high priority.

At Flowerdew Hundred, located in an area where the changeover occurred quite late, archaeology bears out this thesis. The first substantial structures, with full brick cellars, make their appearance sometime toward the middle of the eighteenth century, and virtually all was earthfast construction before that time. So it was that when William Poythress built his one-story frame house with a full English cellar of brick around 1780, he was breaking with a tradition that had been in place at Flowerdew Hundred for a century and a half, and which reached out further in time and space, back to Anglo-Saxon England.

But Morgan's thesis and that proposed by Carson and his colleagues are more compatible than the latter suggested. The settlers seem to have made an initial adjustment to conditions engendered by the tobacco boom, and finding earthfast construction to be a good way to manage resources to maximize profit, they had no reason to change as long as tobacco was king. This brings us in turn to the explanation put forth by Robert Saint George. In examining account books kept by two Connecticut craftsmen, a carpenter and a mason, Saint George was struck by the fact that both men spent a disproportionate amount of time in maintaining their work after it was completed, more than the income from such repairs would suggest. From this and similar examples of other craftsmen, particularly thatchers, he concluded that mutual maintenance was a significant part of a set of reciprocal relationships that lent coherence to a community in preindustrial times. In this light, the earthfast building functioned to maintain social relationships between members of the community. One need not pay for the construction of a structure that would be nearly maintenance free, because in exchange for other services rendered to the house carpenter, repair would be forthcoming at no monetary cost. A kind of "you scratch my back and I'll scratch yours" approach to maintenance existed, so that one was assured of having houses and equipment in reasonable working order with no monetary transactions involved. Does this make sense in a world where labor was costly? It does, for the labor which required capital investment was not that which was needed to repair wagon wheels, patch a thatched roof, or replace a rotted post in a dwelling house. These are occasional needs, met by specific members of the community, not involving a constant day-to-day attending of a crop from which one hopes to make one's fortune. So in a way, Saint George's thesis is but an extension and refinement of that put forth by Carson and his cowriters. Earthfast construction not only makes sense in terms of spending priorities in a situation where most investment must be committed to raising tobacco; it also provides the assurance that one's equipment will be maintained without any significant cash outlay for the service.

Flowerdew Hundred's modest contribution to these dialogues arises from the fact that the changes in question can be observed through archaeology as they occurred in a single location. Support for both the Morgan and Carson theses is provided by the Flowerdew archaeology. The seven sites that were abandoned by the 1660s could well have been those occupied by people driven by a profit motive, producing tobacco until it was no longer highly profitable to do so. The riverside settlement was abandoned at the same time, so the rather abrupt end of a well-defined settlement phase at Flowerdew occurred at the same time as the drop in tobacco prices, thus seeming to uphold Morgan's argument. Yet earthfast building did not stop at Flowerdew Hundred in the 1660s but continued far into the eighteenth century, as did tobacco growing. In this respect, given the correlation between the appearance of the first structures with full brick cellars and a shift away from tobacco cultivation in the Southside, the explanation set forth by Carson and his colleagues seems quite appropriate. Unfortunately, the archaeology provides no insights as to maintenance reciprocity, so Saint George's thesis must remain untested. However, since it does not really contradict the other two views, this is of no great moment. Whether people built earthfast structures because they did not plan to use them for long or because such buildings represented prudent management of limited capital, reciprocal maintenance would work to one's advantage.

The first phase of Flowerdew Hundred's settlement was set into motion by George Yeardley and was given great impetus during Abraham Peirsey's ownership. Following the latter's death in 1628, settlement appears to have leveled off, and during the next three decades it began to diminish. The pipe stem histogram developed by combining samples from all six sites indicates a general end to this first period of settlement ca. 1660, but not all sites were abandoned at precisely the same time. Site 72, the warehouse or barn south of the riverside settlement, appears not to have been in use after 1650, possibly even a bit earlier. The three sites to the south of site 72 appear to have been abandoned progressively later in a southward direction, and the longest-occupied site in the set, site 86, may have lasted well into the sixties. The riverside settlement and the site close by, site 68, both probably lasted until as late as 1660. There is a clear pattern in these dates. The latest locations to be occupied are at the northern and southern end of the line of houses that once extended evenly across the bottomlands, and the ones toward the center were abandoned first. The central reaches of the bottomlands would not be occupied again until the eighteenth century. In the intervening time settlement appears to have remained near the river at the northern and southern ends of the plantation. But it was not a time during which little of moment happened, and the archaeology demonstrates this in a most convincing way.

Chapter Three

By 1700, FLOWERDEW HUNDRED was but a name. Subdivision of the thousand-acre tract began in 1673, and as the eighteenth century passed, it was broken into ever smaller portions. This fragmentation of real estate presents problems for archaeologists and historians alike. As long as there was only one owner, the plantation can be viewed as a single entity, as is the case with Abraham Peirsey. But when numbers of people enter the picture, site locations must be carefully heeded, and the matching of sites with individuals becomes more complex. Tracing changes in title produces an accounting much like biblical "begats," informative but not charged with excitement. But we must account for all of the owners of the property after 1673 if our story is to be complete, and it seems best to do so at this point for the entire time the property was occupied, between 1619 and the present.[1]

It took eight long years for Abraham Peirsey's daughter, Elizabeth Stephens, the rightful heir of Flowerdew, to repatent the land. There were said to be more liabilities than assets when Peirsey's widow presented an inventory to the court in 1628. Land was the major asset, along with tobacco, which had recently fallen in price. Like so many farmers today, Peirsey appears to have been land poor, and the "best Estate . . . ever yett knowen in Virginia" was not the best in financial terms at his death. His daughter restored the name Flowerdew Hundred to the property; perhaps she liked the name better—it does trip from the tongue with greater grace—or perhaps she was asserting her independence by dropping her father's name from the plantation. During at least a part of the eight-year interim, Peirsey's Hundred was under the command of Henry Careless, who may well have lived up to his name in the management of the plantation's affairs. In just three years, Elizabeth sold it to a sea captain and merchant, William Barker. With this sale, the last link to Abraham Peirsey was broken, and while the records are largely silent on the matter, one might guess that Barker was far less committed to the plantation than was Peirsey. He was a prominent citizen, sat in the General Assembly, and was coinvestor in a ship, *Merchant's Hope,* whose name was given to a small Georgian church built in the early eighteenth century which stands nearby today. Tobacco was certainly still being grown, tended by those people living at the location of the line of sites along the Flowerdew bottoms. List making having stopped after 1625, it is not possible to say if these people were those who were living there in 1625 or their descendants; whoever they were, they occupied their homes continuously until the end of the 1650s. Barker died sometime before 1655, and the property passed to his son John. Like his father, John was a businessman, an importer of goods from England to be resold in the colony. John Barker in turn left the plantation to his two sisters, Sarah Lucy and Elizabeth Limbrey. The Lucys and the Limbreys divided the property in half in 1673, the line between northern and southern sections marked still today by a farm road that crosses the property from west to east to the river.

The last two decades of the seventeenth century witnessed an abortive attempt to establish towns throughout the colony, and Flowerdew Hundred was chosen as the site for one of them. The General Assembly's intent in such a program was a more ordered regulation of commerce, with towns serving as marketplaces and only certain designated ports for loading and unloading goods, including tobacco. Each county was to have such a town, and it was hoped that these trade centers would replace the private plantations as prime commercial locations. This effort was but one of a long series of such attempts, going back to 1638. To this end, Charles City County (which then extended south across the James)[2] purchased fifty acres of Flowerdew Hundred for 10,000 pounds of tobacco, and the county surveyor was paid 540 pounds of tobacco for laying out the town. Half-acre lots were offered for sale, on the condition that the purchaser would build on the lot within a period of three months; otherwise, the property rights would revert to the county. Other incentives were offered, including tax relief, particularly if the purchasers agreed not to grow tobacco, which by this time was something of a glut on the market. Al-

12. Seal from wine
bottle, from a trash
pit at site 98

though the records tell us that several dwelling houses and warehouses
were built at the Flowerdew Hundred town, and it was even given the
formal name of Powhatantown in 1702, and market and fair days were
appointed, this seventeenth-century housing development appears to
have been largely a pipe dream. While one or two sites on the property
may have escaped notice, it would be impossible to miss an entire town,
had it been of any significant size. Twenty years of deep plowing by farmers sensitive to site locations and repeated site surveys in the bottomlands
have located all but the most ephemeral of sites. No town has appeared.
Even the town's location is not known with any certainty. One study
places it in the same general area as the early Yeardley-Peirsey settlement
on the river, but other references to its association with a ferry suggest a
location far to the south, where such a service is known to have operated
since the early eighteenth century. Only five of the hundred lots offered
for sale are traceable to named eighteenth-century owners. One of these
men, a wealthy Prince George County merchant, left a trace of his presence, a wine bottle which reads "John Hood, 1751" (fig. 12). This seal was
found in a trash pit on the largest site on the plantation, located at the

southern end of the property near the ferry crossing. So there the matter stands, and only future excavations can add further explanation to the problem of the town that isn't there.

The breakup of Flowerdew Hundred reached its maximum extent during the first quarter of the eighteenth century. Sometime around 1690 John Taylor acquired the northern portion of the property from the Lucys, and in 1707 he divided it among his four daughters, Elizabeth Duke, Sarah Hardyman, Henrietta Hardyman, and Frances Greenhill. In 1702, the southern half was divided into two parcels, each owned by relatives of the Limbrey family, Robert and John Limbrey Wilkins. This partitioning continued in the second half of the eighteenth century, with three people owning portions of the original Limbrey tract: James Warthan on the west and Colin Cocke to the south owned sections, both by 1787, and the third parcel remained in Wilkins hands. While the southern portion was being subdivided, the northern section was being reconsolidated by members of the Poythress family. Beginning with Joshua Poythress the First's acquisition of three of the four parcels owned by the Taylor sisters, a total of 550 acres, the consolidation continued with the addition of 404 acres acquired by Joshua Poythress II, which included some of the acreage owned by Joshua I and the remainder by William Poythress. By 1782 William had taken over all of the southern section except for the southern-most 200 acres, in the hands of Colin Cocke, but he did not hold it for long, selling it to William Peachy, whose widow Mary was forced to sell it to Miles Selden, a Petersburg merchant, to settle her late husband's estate in 1812. Much of this ownership history is a complex web of family relationships, and the majority of the property transfers followed kinship lines. At times when learning who bought or inherited what from whom, one feels a distinct need to have taken a course in social structure. This is particularly so in the case of the final disposition of first the northern portion of the property and eventually the entire 1,000 acres that were first purchased by George Yeardley. It is sufficient to say that the Willcox family, beginning with John Vaughn Willcox who married Susannah Peachy Poythress, owned 735 acres at the north end by 1806 and added parcel after parcel until in 1856 they had put back together that which John Barker and sisters had rent asunder in 1673. From that time onward, the original plantation was to remain intact, passing finally through an intermediate owner to David A. Harrison III.

We must now turn the calendar back to the time following the abandonment of the riverside settlement and the outlying dwelling houses in the bottomlands. There is no doubt that tobacco was still being produced as the main agricultural crop at Flowerdew, but settlement had retreated to the northern and southern shorelines, and the archaeology suggests that people were beginning to look to new ways to make a living. This is the period of the six sites that produced aggregate pipe stem histograms showing a peak toward the end of the seventeenth century and much flatter shapes, the result of longer-term occupation. At least two of these sites were occupied for some time before the earlier ones saw their residents depart, and there is overlap to at least some degree between all seven earlier sites and the six that followed. These were the Barker-Limbrey-Lucy years, the time when Powhatantown was a vision, if not a fact, and they are the least accessible to us today. The records tell us a considerable amount about the day-to-day affairs of the various owners but nothing significant about how their lives were affected by Flowerdew Hundred, or how they affected the plantation in turn. If there is any time period for which archaeology can make a contribution to the story of Flowerdew Hundred, it is this last quarter of the seventeenth century. And indeed, archaeological investigations of two of the sites occupied during the end of the seventeenth century have raised important questions about the colony during these years and have provided at least some preliminary answers, to be built upon through further research.

The two sites in question lie at opposite ends of the property. Site 77, where the large postholes and burials appeared, is southwest of the old Yeardley-Peirsey settlement. It can be assigned to the second half of the seventeenth century by pipe stem evidence and other dated artifacts. Site 92, at the southern end of the plantation, dates to the same period and is the only site at Flowerdew that produced pipe stems of all five bore diameters, although the great majority of them, 72 percent, date to the period between 1650 and 1710. Each site produced a complete English wine bottle datable to the 1650s, the one at site 77 unbroken and the one from site 92 broken but fully restorable, the only two such complete pieces of their kind to have been unearthed at Flowerdew (fig. 13). These bottles have squat, globular bodies and long necks and bear the seals of London taverns on their shoulders. The one from site 92 bears the seal of the Mitre Inn, London, and the initials of William Proctor. The other seal, yet to be identified, shows a crowned head and the initials S. H. Both are good examples of the earliest form of English-made wine bottles of dark green glass, and they provide a dating anchor for the beginning of occupation in the late 1650s or early 1660s.

Neither site revealed an earthfast building when the plow zone was

13. Early English wine bottles, site 77 (*left*) and site 92 (*right*)

removed and the subsoil extensively exposed, but at each, architectural features were encountered quite unlike those found at earlier sites. The feature at site 92 is the more clear-cut and less ambiguous of the two.[3] Excavation revealed a large cellar, sixteen by twenty feet, cut deep in the ground, four feet into subsoil and thus five or more feet below the original ground surface (fig. 14). Around the entire inside periphery of this cellar were large posts set on four-foot centers with walls of planks nailed in place from within. The hole excavated by the builders to receive the cellar was two and a half feet larger in each direction, making it roughly twenty by twenty-five feet in size. The space between the cellar walls and the sides of this larger hole had been intentionally stepped, narrowing abruptly two feet into the subsoil. A wooden partition, still in place, divided the cellar into unequal sections, and wooden stairs led down from

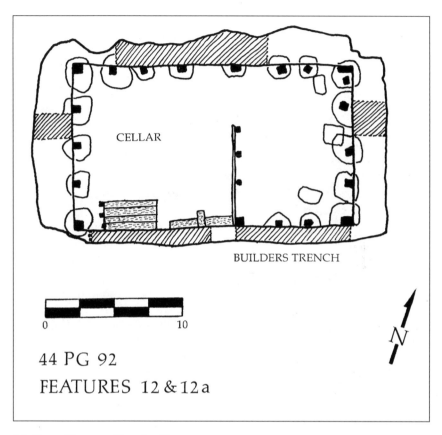

CELLAR

BUILDERS TRENCH

0   10

N

44 PG 92

FEATURES 12 & 12 a

14. Plan of house cellar, site 92

the outside. Sufficient evidence for a chimney was provided by quantities of brick in the fill at one end and in the surrounding area. All of this evidence confronts us with a building unlike anything else seen at Flower-dew, or even in the Chesapeake as a whole. But such buildings apparently were constructed in the seventeenth century, or at least planned, for in 1638 Samuel Symonds of Ipswich, Massachusetts, left instructions for building his house that fit the cellar at site 92 in almost every detail: "I would have the howse stronge in timber, though plaine & well brased. I would have it covered with very good oake-hart inch board, for the present, to be tacked on only for the present, as you tould me. Let the frame begin from the bottom of the cellar & soe in the ordinary way vpright, for I can hereafter (to save the timber within grounde) run vp a thin brick worke without. I thinke it best to have the walls without to be all clapboarded besides the clay walls. It were not amisse to leave a doreway or two within the seller, that soe herafter on may make comings in from without."[4]

In their discussion of earthfast building, Carson and his colleagues referred to the Symonds house as "make do," and thus only an alternative to, but not an improvement over, hole-set buildings. But this may not necessarily be the case, for the "thin brick work without" was clearly meant to protect the "timber within ground," making the building somewhat more permanent. It is by no means a full brick house, yet it is a bit more solid than one framed on posts set into the ground. The stepping of the hole in which the cellar was built is best understood as space for a brick wall, but in this case one that was never to be constructed, for the building lasted but a short time. The cellar was soon filled with what appears to be waterborne sediment, to a depth of a foot or more, effectively preserving the partition in place. The date of the cellar's use, the early 1660s at the latest, and of the postabandonment filling, in the seventies and eighties, leads to the explanation. A hurricane in 1667 did enormous damage to the colony and flooded vast areas of low-lying ground. Site 92 is but a short distance from the river's edge and stands at only nine feet above mean sea level. So we are presented with one of those archaeological rarities, evidence of a natural catastrophe that is known historically and leaves a clear imprint in the record beneath the ground. In this, the filled cellar at site 92 stands in company with Pompeii, entombed by a volcanic eruption, and La Purisima Vieja Mission in California, leveled by an earthquake in 1812 and buried beneath a mud slide. Thus it may be that here, too, thin brick walls were intended but never realized, the "hereafter," to use Symonds's word, never having come. That the space behind the plank walls stood open for some time is proved by cross mends between artifacts from the cellar floor, deposited there while the building was in use, and from the bench formed by the stepping of the sides of the hole.

So what are we to make of a house of this size, quite small by twentieth-century standards? In the first place, it was not all that small for its time. Most planters, even those of the middling sort, lived in one-room dwellings of the same size, a fact often overlooked due to the size of larger surviving buildings. It would be enormously helpful if we could link this building to one of the cast of Flowerdew Hundred characters. John Barker or Robert and Elizabeth Limbrey are possible candidates, but it is not certain whether John Barker ever resided at Flowerdew, and the Limbreys would have had to have been living on the site before they acquired the land in 1673. Since Elizabeth was John Barker's sister, this latter possibility cannot be ruled out. But whoever was responsible for erecting this unusual dwelling house, that person seems to have been making a statement in wood and brick, a statement of commitment to place, however tentative it may have been.

Much has been written in recent years about the way people use their material culture—houses, clothing, dishes, food, and so on—to set themselves apart from others as a distinct group. While there can be no question that this happens, many of the explanations for such group identity have placed undue emphasis on "power relationships," the statement in material things by the members of a ruling elite to declare their position of dominance. There is no question that a great brick Georgian mansion, such as the one at Westover Plantation across the river from Flowerdew, does just that. The house at Westover affirmed its owner William Byrd's position in the community, as well as in the smaller precincts of slave quarters and other modest dwelling houses on the plantation. But power relationships are not the only things that are communicated by having a different house or eating different food. We are all familiar with how various ethnic groups in America today use material culture as a means of maintaining their ethnic identity, and even various subcultures of mainstream America do likewise. Bikers, cowboys, and even some good ol' boys follow suit, as witness the title of a country classic, "Red Necks, White Socks, and Blue Ribbon Beer." So the essence of material culture in its relationship to group identity lies in differences which can be perceived, and some of these differences might relate to superior-inferior relationships, but by no means all. There is considerable debate, as well, as to what degree such "group boundary marking" is done consciously.

Good judgment suggests that few if any people agree before the fact to let this kind of house or that type of dress identify them as members of a group. Rather, they adopt certain practices that become significant after the fact and are just as potent. It is against these considerations that we can view the little cellar house at site 92 in a new light. Those responsible for its construction may or may not have been of the emerging elite of the late seventeenth century. But one thing is certain, the house is quite different from its contemporary earthfast buildings.

Before discussing this difference and the possible causes behind it, we must pay a brief visit to the other location, site 77, across the peninsula on the river to the north. An area of some 2,500 square feet was cleared and carefully examined for postholes. A large cellar was encountered in the center, and postholes appeared in profusion, including the line of large ones that might conceivably be those of Peirsey's rail. The only problem was that the postholes steadfastly refused to provide a pattern that would even remotely suggest an earthfast house. We are left with only a cellar, large and floored with square clay tiles. Fifteen feet long, eight feet wide, and four feet deep, the cellar is too small to have held the framing of a house as at site 92, but on the other hand, it is larger than most root cellars found with hole-set houses. It is possible that the building that once stood on the site was erected on ground sills, in which case plowing would have obliterated all evidence of the frame, but this is only conjecture. Nevertheless, site 77 is still significant and relates to our story, not for what it was, but what it wasn't. Like the house at site 92, it was not an earthfast building, thus setting it apart from those that precede it and others contemporary with it at Flowerdew Hundred. The cellar fill yielded a rich collection of artifacts from the second half of the seventeenth century, including large numbers of locally made clay smoking pipes, many of them elaborately decorated with a variety of geometric and naturalistic designs. As with site 92, we do not know who built or lived in the house that once stood over the tiled cellar. Perhaps it was John Barker, it might have been the Lucys, and it is equally possible that it was someone whose name we will never know.

The other four sites in the group which includes sites 77 and 92 have yet to be excavated, but their location is significant, for with the other two they form two sets of sites, three in each, at opposite ends of the plantation. From the different settlement pattern of these six sites, and the architectural information that the excavated ones provides, it is possible to suggest that their occupants might have been involved in something other than the production of tobacco. Other evidence from site 92 points strongly in this direction.

To the south of the house at site 92 were three irregular shallow pits and the remains of a small hole-set structure (fig. 15). The pits were not the typical trash pits of the period: the artifact content was very low, and the major part of the fill consisted of charcoal, slag, brick, baked clay, oyster shell, coral, and iron. Small stake holes surrounded each pit, indicating some kind of enclosure or screen. When the surface collection was made in the area above these pits, lead ore fragments and a crucible base were found. Taken together, this is ample evidence for some kind of in-

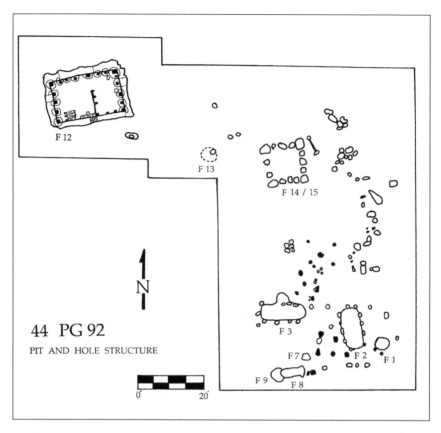

F 12

F 13

F 14 / 15

N

44 PG 92

PIT AND HOLE STRUCTURE

F 3

F 7

F 2

F 1

F 9

F 8

0   20'

15. Plan of site 92, showing house cellar, hole-set outbuilding, and roasting pits

dustrial production having taken place at the site. The most likely activity, in spite of certain ambiguities in the evidence, would be the manufacture of small quantities of low-grade iron, using the bloomery method of production. Iron smelting using a blast furnace never took place in seventeenth-century Virginia. The company began building such a furnace at Falling Creek, near modern Richmond, but the 1622 uprising destroyed it, and no further attempts were made. This is in sharp contrast with New England, where a productive blast furnace was established at Saugus, Massachusetts, in 1640.

This difference may account for another difference seen in the archaeology of the two areas. Probate records, detailed lists of property made for tax purposes at the time people died, show that armor was as common in

New England during the second quarter of the seventeenth century as it was in Virginia. But while Virginia sites of the period, Flowerdew Hundred among them, have yielded great quantities of discarded armor, including two spectacular sixteenth-century close helmets from Martin's Hundred, only the odd scrap of armor has been found on New England sites. Although not as many early sites have been excavated in New England as in Virginia, the difference is striking nonetheless. Wholesale disposal of armor in Virginia can be understood simply by imagining oneself clad in steel in the July heat. What's more, the armor recovered from these Chesapeake sites was obsolete when it arrived in the colony, war surplus as it were, leftovers from the stores in the Tower of London. The colonists appear to have had little use for it (although it is carefully noted in the 1625 muster, but then so is fish) and threw it in the nearest handy hog wallow or open trash pit. With no means to recycle such cumbersome equipment, there really was little choice. In New England, however, the advanced ironworking technology made recycling armor possible, particularly during the Puritan Revolution and the Commonwealth period when supplies from the mother country were uncertain.

But for Virginians who wished to produce iron, the only means at hand was bloomeries, and evidence for these has turned up on at least three sites. We know that it was being done in the colony, and site 92 suggests its presence at Flowerdew Hundred.[5] Bloomeries produce iron from a low-grade ore known as bog iron, deposits of which are numerous in the Flowerdew area, including one on the plantation. The technology is simple and archaic, going back to the European Iron Age, a millennium before Christ, and even longer in the Middle East. It is also the technology used by African ironsmiths today. The bog iron is first roasted in shallow pits, to render it more friable, and then placed in a simple clay-lined furnace with charcoal and lime. The lime acts as a flux and reduces the temperature at which the process will occur. The result is a spongy mass of slag and iron known as a bloom, which can then be hammered into bars for further use. Such a technology, small in scale and simple in execution, is ideally suited to modest production to meet immediate needs and can be carried out by only one or two people if need be. While there are certain ambiguities in the interpretation of the pits at site 92—for instance, samples submitted for analysis did not exhibit quite the ore richness expected—the evidence is compelling nonetheless. The stake holes surrounding the pits recall the screened pits at bloomeries shown in sixteenth-century graphics, and the baked clay could have lined the simple furnace. Shell and coral produce lime; there were hematite nodules in the pits; and the quantities of charcoal are large. In a way, the pits are like spoor to the hunter, who can see the tracks which prove a lion is close by, even though the animal is not in sight. Likewise in archaeology; the "smoking gun" in this case may well be located beneath rows of tall corn, awaiting discovery.

What was the relationship between the people who lived in the cellar house and the industrial activities taking place in the adjacent yard? Ceramic cross mends between the fill of the pits and the upper postabandonment level of the house suggest that the house predates whatever was happening in the yard. But other evidence points toward their being contemporary. Of the 142 fragments of glass bottles found in all three pits, 127 are of case bottles, and only 15 of dark green English wine bottles. Case bottles, made both in England and Holland between 1625 and 1675, are rectangular in section, with short necks; they fit snugly together in wooden cases. English wine bottles are globular, with longer necks, and were manufactured from the 1650s onward. Although case bottles were made up to the end of the third quarter of the seventeenth century, their temporal distribution on Flowerdew Hundred sites is largely limited to the first half of the century. In view of this, and the small number of English wine bottle fragments from the pit's fill, a date in the early 1660s does not seem unreasonable. The principle of *terminus post quem,* that is, a deposit must be dated by the most recent object found in it, is a sound one when applied to discrete deposits that occurred in a single brief filling episode. But the cleaning up of an area, moving debris from one place to another, at times can render the principle less than precise. The house could have been occupied and the pits in use at the same time, and later, pieces of the same pot could have been dumped in the abandoned cellar and also into the pits when they were topped off and leveled, no longer in use.

Such abundant if somewhat ambiguous evidence of an emerging local industry is lacking at site 77. But a single artifact found there raises the possibility that another commodity may have been produced there. The object is a "waster" of a locally made clay smoking pipe. These pipes, large numbers of which came from the cellar fill at the site, are easily distinguished from their English-made counterparts. The clay used in their manufacture is of a darker color than the stark white of the English pipes, almost a terra-cotta color. Even those that are undecorated can be recognized at once as homemade products. Their stem bore diameters are considerably larger than those of English-made pipes, and thus they cannot be used in dating. Wasters, in ceramics lingo, are malformed pots or pipes, misshapen either through poor potting, an accident of firing, or oversight on the maker's part. The pipe fragment in question, part of a bowl with a bit of stem attached, was probably the result of an oversight, for the stem was never bored. It is easy to imagine how such a thing could happen. Pipes of this sort were turned out in large numbers at one time, and if a number of molds were in use at once, one could have been overlooked, and the pipe taken out without the wire ever having passed through the stem. The oversight would probably not have been noticed until after firing, as was the case with this waster from site 77. This particular pipe-making mishap is especially relevant to the question at hand. Some wasters, such as those with misshapen bowls or bent stems, were still usable, much like factory seconds today. But a pipe without a bore would be of no use whatsoever, unless there were people given to practical jokes in the seventeenth century who would give such a pipe as a gift. Its particular deformity suggests that it was discarded near where it had been made.

How then does all of this talk of local industry, permanent and not so permanent building, ethnic identity, and Flowerdew Hundred relate to larger historical issues? The answer is to be found by considering the way in which seventeenth-century English people eventually became Americans and the time when this great transformation occurred. My book *In Small Things Forgotten* proposes a three-phase development of the culture of colonial America away from that of parent England, based on the archaeological record as well as the written word. Colonial material culture in America between 1607 and ca. 1660 appears much like that which its makers and users knew in England. Variations in building styles reflect the demographic diversity of the early settlers, rather than solutions to various problems presented them in the New World. This period, roughly two generations long, witnessed the first native-born generation coming to majority.

A degree of isolation from England during the Puritan Revolution and the Commonwealth led to these "new Americans" becoming less and less English during the second phase, between 1660 and 1760. Even though interest in and the regulation of the affairs of the colonies was vigorously renewed under Charles II, the vector of change had been set, and it took until the eve of the American Revolution for colonial culture to become more like that of England again. During these hundred years, distinctive styles of building developed that were expressions in wood, brick, and clay of American ideals and attitudes, not those of the mother country, and these also exhibit a great degree of variety, the result of geographical isolation, a conservative folk tradition, and different environmental constraints. This is particularly so in New England, where by the end of the seventeenth century, building techniques in Massachusetts, the Old Colony of Plymouth, Rhode Island, and Connecticut all show important differences from one another. Other aspects of material culture reflect the same changes during the period. Patterns of ceramic use diverge from those of England, and the elaborate designs carved atop gravestones are unmistakably American, spirit faces that were never to be seen on the opposite side of the Atlantic.

The third phase of this development represents a major break from the two that preceded it. Whether English or Anglo-American, people in the colonies, outside the large urban areas such as Boston, Philadelphia, or Charleston, lived in what is commonly called a folk society, one that placed great value on small-scale community life, extensive social networks that placed the group ahead of the individual in importance, and one that was integrated through a set of reciprocal relationships. It is in such a world that maintenance relations as described by Saint George flourish and foster community solidarity. As the eighteenth century came to an end, all of this had changed, to be replaced by a world in which the individual took precedence over the community, where privacy became increasingly prized, and where social mobility became more a matter of individual initiative. The reasons for this great transformation relate to the passing of a world in which one's place was essentially fixed, in this life by the Elizabethan version of the Great Chain of Being which ordered the universe in a strict hierarchy with God at the top and inanimate objects at the bottom, and in the next by Calvinistic predestination. But with the development of a more secular world and the impact of the thinking of people like Copernicus and Newton, the passage to the modern world had taken place.

In *Folk Housing in Middle Virginia*, Henry Glassie detailed this transformation as shown by vernacular buildings in Louisa County in the Virginia piedmont. There, as elsewhere in the colonial world, the old hall-and-parlor house gave way on the landscape to a new house type, a central-hall I house, so called since they are very common in three states which begin with the letter I, Iowa, Indiana, and Illinois.[6] Derived from the severely symmetrical Georgian house form, with two rooms front and back flanking a central hall, these frame houses are only one room deep but retain the hall in the center. They are private where the hall-and-parlor house is public; entry does not lead directly into the bustle of the hall but rather into a hall from which the inner parts of the house are reached. These I houses were not larger than the hall-and-parlor dwellings that preceded them, but they divide the same volume into more private spaces. In the case of a hall-and-parlor house, the interior arrangement was imaginable from the outside; one could tell where food was being cooked, or where the master of the house would spend the night. Not so with an I house, which presents a blank facade to the world, balanced, ordered, but mute as to what is taking place behind the doors.

The changes Glassie demonstrated in the folk building of the Virginia piedmont can be seen to occur in most aspects of material culture over all of colonial America. The appearance of sets of matched dinner plates at the same time speaks to individual service rather than the earlier pattern of sharing food containers. Even the manner of serving the food changed, from stews and other mixed dishes to individual portions, often served separately on the plate, much as we do today. Chests of drawers, in which the contents were segregated, replaced lidded chests, all-purpose containers which could hold one's possessions packed together in a single unpartitioned space. Epitaphs on gravestones changed from public first person—"Remember me as you pass by"—to private third person—"His soul has gone to heaven"—and even the disposal of refuse changed from broadcasting it over the landscape or putting it into shallow holes to disposal in carefully constructed trash pits, made especially for the purpose. If the expression "A place for everything and everything in its place" did not have its origin at this time, it certainly aptly describes the attitude that the material culture, from houses to dishes, indicates. Order, control, individualism, and privacy ruled people's behavior and shaped the way the world was perceived.

It is against the background of these sweeping changes in American colonial life that we can begin to make some sense out of what was happening at Flowerdew Hundred in the late seventeenth and eighteenth centuries. The cellar house and adjoining pits at site 92 represent in microcosm changes in the world at large and one person's response to them. Slim though the evidence may seem, it is compelling nonetheless when set into a broader context. Taken alone, neither the evidence for industrial activity involving iron in some way nor the odd little house framed within its cellar is all that telling, but together they give us a picture of the way in which some people in the Chesapeake were organizing their world during the second phase of Anglo-American cultural development and change.

In Small Things Forgotten was published in 1977, before the important scholarship on earthfast building and the sites that produced its evidence had made its appearance. What we know now about earthfast, impermanent building and its economic implications provides a valuable body of comparative material to supplement that on which the book was primarily based, which had a decided New England bias. New England and Virginia present us with a study in contrasts. New England's economic diversity developed quite early and included iron production, shipbuilding, mixed-crop farming, timber industries, fishing, and even whaling before the end of the seventeenth century. From the start the northern colonies were established by families, arriving in great numbers, and unlike that of fever-ridden Virginia, New England's climate was most salubrious. People there actually could expect to live longer than if they had stayed in England. Such a combination of circumstances favored strong economic development almost from the beginning. So it is that earthfast building appears rarely, with only three known examples from archaeology and another two from the documents, including a trading post at Manomet in Plymouth colony which was destroyed by a hurricane in 1635, as we are told by William Bradford: "It began in the morning, a little before day, and grue not by degrees, but came with violence in the begining, to the great amasmente of many. It blew downe sundry houses, and uncovered others; diverce vessells were lost at sea, and many more in extreme danger. It caused the sea to swell (to the southward of this place) above 20 foote, right up and downe, and made many of the Indeans to clime into trees for their saftie; it tooke of the borded roofe of a house which belonged to this plantation at Manamet, and floted it to another place, the posts still standing in the ground; and if it had continued long without the shifting of the wind, it is like it would have drouned some parte of the cuntrie."[7]

Fully framed sturdy buildings, which appear in New England almost from the start, are a statement of commitment to place as well as the result of an economy that permits capital investment of substantial amounts in such construction. Virginia by contrast became locked into a single economy from the time John Rolfe introduced Oronoco tobacco in 1610–11. And so it went, right through until the 1660s, when the market for tobacco went into a serious decline, making it desirable, if not necessary, for at least some people to look to other ways of making a living in the colony. Becoming American almost by definition involves achieving some kind of independence from England. More than a century before obtaining political independence, people in the colonies were developing various domestic industrial enterprises which would lead them if not to economic independence, at least away from total reliance on England.

That this change was taking place is evident not only in what we know of the development of various industries but also in the way the crown reacted to this development. Fearing competition for its own products and already having passed the Navigation Acts which limited colonial consumption to that of English-made goods, Parliament enacted other legislation against the local production of needed commodities. At the same time the Virginia assembly was promoting local manufactures through legislation of its own. During the second half of the seventeenth century, a variety of domestically produced goods made their appearance, including shoes from locally tanned leather, cloth, smoking pipes, and even ships. Ceramics in the form of undecorated coarse utility wares, such as milk pans, butter pots, and storage jars, were also produced in Virginia from quite early on; Noël Hume has unearthed evidence of pottery production in the 1630s at Martin's Hundred. But these should probably not be included in this inventory of local industries, for production of pottery began in New England at about the same date, and these plain wares were never perceived as a threat to the thriving ceramic industry in Staffordshire. They have such low values assigned to them in probate inventories that it probably was not economically feasible to export any more kitchen wares than necessary. In fact, significantly less locally produced pottery is found on Virginia sites than on contemporary ones in New England, suggesting that in this single category New England had achieved a greater degree of independence than had Virginia, which was actually more dependent on England for its ceramic needs.

Whether the manufacturers of these locally produced commodities formed an elite is highly questionable. Certainly the true elite that arose in Virginia during the eighteenth century was based not on industrial production but rather on a new economy of tobacco produced by slave labor. Robert Carter, the richest man of his time in the colony, was a wealthy eighteenth-century tobacco planter and nothing more, his fortune produced by the labor of hundreds of Africans who worked his fields. The industrial production indicated by the Flowerdew Hundred evidence seems tentative and experimental, but we must not let this detract from its significance. And clearly the person who built the house at site 92 knew that he was engaged in something very different from what occupied his neighbors' time. While the house signaled more hope than realization, it is the intent that counts, the desire for a house that would be different from the hole-set ones to the north. Had the brickwork run above ground level, the house would have given the appearance of one on solid brick footings, setting its occupants apart from those living in other houses on the plantation. It is altogether possible that the attempt at producing iron and making implements from it was the work of John Barker, even if he did not reside at Flowerdew Hundred himself. A merchant who sold imported English goods, he may well have been quick to realize the benefits of producing some of these himself and set about to have this hope realized. The same possibility could apply to site 77, if pipes were in fact being produced there, although admittedly the evidence is quite tenuous.

So we can suggest that the architectural diversity during the second half of the seventeenth and first half of the eighteenth century in Virginia is nearly equal to that of New England, but that the reasons for it are quite different. New England, with a highly diversified economy from the start, developed differing types of buildings as a function of regionalism, and the regions reflect in part boundaries between colonies. Virginia on the other hand established an English form of building, though one that was fast disappearing at the time, and maintained it through the seventeenth century and into the eighteenth in some areas. Diversity there is the result of a total commitment to one economy until the second half of the seventeenth century, when other forms of building made their appearance among a certain group of people who had renounced this commitment in favor of something else. In so doing, they achieved a degree of economic independence, not only for themselves but for the colony as a whole, and, more importantly, a cultural independence as well. So in the two regions, becoming American took two different courses, but with similar end results.

William Kelso has shown that in Virginia house forms remained more English than in New England.[8] He attributed this to greater numbers of English people arriving in the Chesapeake than in the northern colonies throughout the seventeenth century. Furthermore, we have seen that the strong commitment to tobacco brought about the retention of earthfast buildings into the eighteenth century in some areas. And while Kelso saw a three-phase pattern of change like that described in *In Small Things Forgotten,* it appears to take place somewhat later. If Flowerdew Hundred can be seen as a microcosm of the Chesapeake region, these conclusions are borne out by the archaeology there. Full brick cellars for frame houses do not make their appearance until the 1740s, and these buildings were almost certainly examples of the so-called Virginia house with end chimneys, either of a single room or two-room, hall-and-parlor plan. More Georgian-influenced house forms do not appear to have been built at Flowerdew Hundred until the final quarter of the eighteenth century, on the basis of the archaeology as it is presently known. And since the change to mixed-crop farming took place at different times in various regions of the Chesapeake, at any given time after the third quarter of the seventeenth century we can see considerable regional diversity in house form, the result of differing economies, not only agricultural but industrial as well. This is most certainly the case at Flowerdew Hundred, not only at sites 77 and 92 but also at site 98, which has the earliest house with a full brick cellar. This house appears to have been connected with yet another commercial enterprise, maintaining a ferry service connecting the road between Williamsburg and Petersburg.

As the century came to a close, the slave ships began to arrive in increasingly greater numbers at Virginia ports, setting the stage for the development of a culture more American than anything that had gone before. The seeds of this new order were planted sometime before slavery became the foundation for the Virginia way of life. Dell Upton has told us that the roots of the classic plantation settlement pattern, the master's house with slave quarters set at some remove, lie in growing conflict between masters and indentured servants as early as the 1660s.[9] His conclusions were based on a study of room-by-room probate inventories, those that list a house's contents according to the room in which they were located, taken between 1640 and 1720. As one would expect, most of these inventories are of small houses, one to three rooms in size. But when one group of houses is considered, those of eight to eleven rooms, a surprising pattern of change over time emerges. Houses of this size increase in numbers peaking in 1680 but then become fewer, falling back to their 1640 number by 1720.

The explanation offered for this rather curious pattern of growth and decline relates to social relationships between masters and servants. Until the 1660s it was accepted practice for masters and servants to occupy the same house, so that the increase in numbers of large houses reflects the growing numbers of people who could afford servants. But increasing conflict between masters and servants, beginning in the 1660s, led to new arrangements of living space. This conflict is witnessed by a sharp increase in servant-master litigation in the court records, servants claiming a breach of contract and masters feeling they were not receiving the service to which they were entitled. The masters reacted to this mounting conflict by removing the servants from their quarters within the house and installing them in separate buildings at some distance away. This then accounts for the reduction in the number of large houses that occurred after 1680. A contemporary description of these late seventeenth-century plantation communities makes the point quite clearly: "Some people in this country are comfortably housed. . . . Whatever their rank, and I know not why, they build only two rooms with some closets on the ground floor, and two rooms in the attic above; but they build several like this, according to their means. They build also a separate kitchen, a separate house for the Christian slaves, some for the negro slaves, and several to dry the tobacco, so that when you come to the home of a person of some means, you think you are entering a fairly large village."[10]

The timing of this change makes it but another example of the divergence of American society from that of the England from which it came. As fully institutionalized slavery came to form the foundation of colonial Virginia plantation life, this way of laying out a settlement would have an important bearing on the formation of the archaeological record, in one of its most intriguing and controversial aspects, and Flowerdew Hundred plays a role in the explanation of that record.

Chapter Four

A REMARKABLE PIECE OF earth sculpture was created by archaeologists excavating an abandoned cellar on a site at the northern end of Flowerdew Hundred. "Pedestaling," an accepted technique in archaeological excavation, involves the intentional retention of an artifact's position in the ground by leaving it on a pedestal of earth, as the surrounding fill is removed. In most cases such pedestals are but a few inches in height and permit the observation of the in situ relationship between groups of objects. But the archaeologists working in this cellar seemed to have been intent on something more. The fill of the cellar was very rich, and numbers of complete or nearly whole bottles, ceramics, and metal tools were encountered. Someone decided to try to pedestal almost everything, and since the deposit was quite deep, some objects wound up on earth pillars more than a foot high. The result was a spectacular aesthetic achievement and showed a group of typical early eighteenth-century objects in their precise position in the cellar fill. On the other hand, not much knowledge was gained from this exercise in carving the soil. It is quite clear that the cellar filled in a random way, and the relationships between the objects in it was the result of chance. Nonetheless, the visual effect is quite striking, and the deposit has been painstakingly reproduced in the museum at Flowerdew Hundred. But other artifacts from the cellar and a nearby well have an important story to tell, leading us into one of the most lively controversies in Chesapeake archaeology. To understand this, we must place the site, number 66, in both its Flowerdew Hundred setting and that of the Chesapeake region as a whole.

Site 66 is one of the five that make up the third set of sites representing the occupation of the Flowerdew bottoms between 1619 and the middle of the eighteenth century. All five show pipe stem histograms that peak before 1750, and like the first group of sites, they are located across the bottomland at rather equal intervals, occupation having moved once again into the central portion away from the river. The drop in the histograms during the period between 1750 and 1800 marks a shift in the overall settlement layout at Flowerdew Hundred that involved the removal of almost all buildings from the fertile bottomland fields and the construction of new ones on a ridge to the west. This relocation was almost certainly the result of changes in means of transportation during the early eighteenth century. The river had been the main avenue of travel and transport before the development of an adequate system of roads, so locating one's house lot in reasonable proximity to the James was a logical thing to do. But once there were roads suitable for travel, the rich bottomland soil could be given over entirely to cultivation. The ferry that was established in the early years of the eighteenth century suggests that roads were being developed by that time, for a ferry makes little sense unless it serves as a connection between roads on opposite sides of the river. By the time of the Revolution, the only structures remaining in the bottomlands were at the southern end of the plantation, and these appear to have been connected with the ferry operating there.

The initial occupation of these five sites occurred at the time when slavery became a formal institution, based strictly on race, and the numbers of Africans coming to Virginia rose in spectacular fashion. This relationship must be more than coincidental, and when materials from the sites are taken into account, there seems little question at all. Site 66 produced quantities of a very distinctive kind of ceramics, as did the surface collections from the other four sites in the group; however, none has been recovered from sites in the earlier two sets. The pottery is quite different from all other types found at Flowerdew Hundred, those either of European manufacture or made locally by colonial potters. It is an unglazed, handmade ware, tan to dark gray in color and fired at a rather low temperature, probably simply by stacking the pots with fuel and burning them in the open. The commonest form made in this ware is a shallow bowl with a flattened rim and flat base; in the Chesapeake other shapes include

16. Colono ware chamber pot. (Courtesy of the Colonial Williamsburg Foundation)

skilled copies of English forms, including milk pans, porringers complete with pierced handles typical of pewter examples, pipkins (cooking pots with three legs), large punch bowls, chamber pots, and even teapots (fig. 16). Archaeologists have given two names to this pottery, Colono-Indian ware and just plain Colono ware. The two names reflect a deep difference of opinion about exactly who was responsible for producing this pottery. Using the less specific term *Colono,* we must consider this controversy in some detail here, for the identity of the makers of these pots has important implications about the nature of Chesapeake society during the seventeenth and eighteenth centuries. The discussion leads us far from Flowerdew Hundred and then requires that we take a closer look at those curiously decorated smoking pipes of the kind found in the cellar at site 77 and on other sites of the later seventeenth century on the plantation.

The basic question is simple. Was Colono ware made by slaves or, as the other name implies, by native Americans? By its very nature, there can be no final, definitive answer to this query. Yet the material from Flowerdew Hundred, even when taken alone—the pipe stem histograms and surface surveys and the way these bits of evidence are accommodated by the historical record—points strongly to slaves as the makers of this pottery. With the record of two centuries of occupation assembled at a single location, such an exercise becomes possible.

Long before it was identified in formal terms, Colono ware had been turning up in excavations of colonial sites in Virginia. It was first encountered at Williamsburg in the 1930s and at Jamestown at about the same time. Four eighteenth-century sites in Yorktown produced it between 1957 and 1962. But not until the latter date was its existence explicitly pointed out by Ivor Noël Hume in a brief article in the *Bulletin* of the Virginia Archaeological Society.[1] The argument advanced in this article is logical and consistent but seems almost to assume as a given what it sets out to prove. Put another way, even the slightest possibility of people other than Indians making the pottery is not acknowledged, and an explanation for its pattern of occurrence is developed assuming Indian manufacture from the outset. Calling it "an Indian product manufactured during the colonial period," Noël Hume went on to say: "As a generalization, it may be claimed that most excavated colonial sites in the vicinity of the James and York Rivers from which records or artifacts survive have yielded fragments of this Indian ware. It would be logical, therefore, to assume that it was manufactured by local Indians who were exposed to European influences, presumably on the reservations of the Mattaponi, the Chickahominy or the Pamunky."[2]

At the time no archaeological work had been done at either the Mattaponi or Chickahominy reservations, leaving only Frank Speck's 1928 study of Pamunky to attest to the presence of such pottery at sites known to have been occupied by Indians at the time.[3] But when we look at Speck's accounting of his work, the first thing that stands out is that none of the pottery he found there, which he described in general terms as something that resembles Colono ware, has a securely dated context. On the basis of this rather shaky evidence, Noël Hume concluded that "the Pamunky Indians manufactured the examples of the ware encountered on colonial sites in the Gloucester, Yorktown, Jamestown area." He explained their presence through the eighteenth century as the result of planters acquiring pottery from the Indians for use by slaves, "being loath to purchase fancy English pottery" for them. But one must ask why it had to be "fancy"; given the low price of locally produced coarse wares, these would have sufficed just as well, and the supply would have been more assured. Colono ware appears to date from ca. 1680 through the last quarter of the eighteenth century, becoming more abundant with the passage of time. Its near disappearance at the end of that century probably was due to the increasing availability of inexpensive Staffordshire pottery, for it is this that we find in quantities in excavated slave quarters from the earlier years of the nineteenth century.

Three years after Noël Hume's initial description of Colono ware appeared, Lewis Binford, writing in the same journal, described pottery that he had recovered from five sites in southeastern Virginia.[4] All of the pottery was collected from the surface, and since no other details are given regarding other material from the sites, it is not possible to determine just what the sites represent or how long they were occupied. Pipe stems were also recovered and dated using a formula which Binford himself developed that gives a mean date but no indication of the duration of occupation such as one obtains using Harrington-type histograms. The mean date in this case was 1675, but since there is no way to tell how long the occupation lasted before and after that time, we are somewhat at a loss. In any case, 1675 is only a few years earlier than the earliest dates known for Colono ware from other Virginia sites. There is also a problem with the identity of the occupants of the sites. Binford attributed them to two Virginia Indian groups, the Weanock and Nottoway, but the Weanock withdrew from the area in the period between 1653 and 1666, earlier than the mean date suggested by the pipe stems, while the Nottoway remained into the eighteenth century. Binford is cited in an article by Leland Ferguson as having stated that the sites in question were in fact occupied by both Africans and Indians.[5] Commenting on the flat bottoms of the vessels, Binford suggested that these might reflect cooking on an English-type hearth rather than an open fire. Indian pots do not share in this characteristic but have rounded bases better suited to standing in coals and ash. If this is the case, do the flat bottoms "betray a major change in house types among the Indians?" One can equally well ask if they were used, if not made, by people who lived in English-style houses, as we know slaves did.

As more and more Colono ware appeared in sites around the region, its identification as the product of Indian workmanship rested solely on these two articles, neither of which presents conclusive evidence. Both rely on Speck's material from the Pamunky Reservation, which is also open to question regarding its date. In view of the fact that Colono ware continued to be produced through the end of the eighteenth century, and in a few places has been found in early nineteenth-century contexts,

Thomas Jefferson's comments on the Pamunky written in 1781 take on special significance:

> Chickahominies removed, about the year 1661 to Mattapony River. . . .
> They retained however their separate name so late as 1705 and were at
> length blended with the Pamunkies and the Mattaponies and exist at
> present only under their names. There remain of the Mattaponies three
> or four men only, and they have more negro than Indian blood in them.
> They have lost their language, have reduced themselves, by voluntary
> sale, to about fifty acres of land, which lie on the river of their own name,
> and have, from time to time, been joining the Pamunkies, from whom
> they are distant but ten miles. The Pamunkies are reduced to about ten
> or twelve men, tolerably pure from mixture with other colours. The older
> ones among them preserve their language in a small degree, which are the
> last vestiges on earth, as far as we know, of the Powhatan language. . . . Of
> the Nottoways, not a male is left. A few women constitute the remains of
> that tribe.[6]

This state of affairs was but the end point of a rapid drop in the Indian population in Virginia that was already in full course more than a century before. When the assembly enumerated the Indian males in the colony in 1669, their number had already shrunk to 725, representing nineteen tribes. It seems rather curious that as the number of Indians dropped, the quantity of Colono ware was moving in the opposite direction, closely matching the increase in the colony's slave population.

Colono ware was first thought to be a strictly local product, found only on Chesapeake sites. But the matter turned out to be more complex, for we now know that it has a continuous distribution throughout the Old South, from Maryland to Georgia and west into Tennessee. While exhibiting some variation, particularly in shape, it is all the same pottery, and the variations in fact provide a basis for a convincing explanation of its production and the identity of its makers. This wider distribution was just being made known when Noël Hume published his 1962 article, and he made reference to its having been found by Stanley South of the University of South Carolina on three North Carolina sites as well as in Charleston, South Carolina. South was the first to question the attribution of Colono ware to Indian makers, and he was to be followed by other researchers during the 1970s. It is significant that the first doubts were raised by archaeologists not working in the Chesapeake, for it is one thing to suggest that the Pamunky and other fragmented Virginia groups were making the pottery but quite another to see them as responsible for a product distributed over the entire colonial South.

Finally, in 1977, Richard Polhemus stated what now seems obvious, that Africans and not Indians were responsible for the manufacture of the pottery. In his report on the Tellico blockhouse, an early nineteenth-century site on the Little Tennessee River, Tennessee, he commented, "The single historically documented factor linking 'Colono-Indian' pottery producing settlements on the coastal plain of Virginia, and North Carolina, with those in the uplands of South Carolina and the Tellico blockhouse, is the presence of Negroes at the sites in question." Polhemus apparently had been thinking about the problem for some time, for in 1974 he studied pottery from Nigeria and Ghana and found that it was very similar to Colono ware from South Carolina sites. Although the African pieces were not dated, certain of the Ghanaian pots were absolutely indistinguishable from the South Carolina ones. Both were flat-bottomed, burnished, and grit-tempered and had incised X's on their bases.[7]

Leland Ferguson, a colleague of both South and Polhemus, commented in 1980 that "with this important observation, the lid was cracked on a box that has sat covered with dust in the darkest corner of North American historic sites archaeology, the contribution of Afro-Americans to the pottery we call Colono-Indian." It was Ferguson who first proposed the term *Colono* in place of Colono-Indian, and in his article "Looking for the 'Afro' in Colono-Indian Pottery," he set out a well-reasoned argument for the identification he proposed. He suggested that two pairs of questions should be asked: who made it and when, and who used it and how was it selected? It is possible that all three groups—Indians, Africans, and Europeans—made it, but we can almost certainly rule out the Europeans, for they were possessed of a technology of a very different sort, and even the simplest European pottery is wheel made. This leaves Indians, Africans, or possibly both. Ferguson cited "clear and well-documented evidence" that Indians had in the past made pottery that might fit the general category in being handmade, burnished, and unglazed. But the only group in the South Carolina region who made pottery of this kind were the Catawba, who made items of this type for sale in the nineteenth century. There is but a single documentary reference to a late eighteenth-century sale of pottery by the Catawba. Although Ferguson made no mention of it, the Pamunky were producing similar pottery in the nineteenth century as well, and this could well be the ware that Speck found on their reservation. Ferguson went on to say that some Gulf Coast tribes, notably the Natchez, made pots in European forms, but there the resemblance stops, for they are still decorated in traditional methods, paddle stamping and incising. But this is so different that it does not fit even Noël Hume's original definition. Colono ware from Virginia to Georgia is essentially similar in manufacturing technique, though there are the different shapes, and shares equally both African and Indian pottery traits that are too general to be in any way diagnostic, all three being handmade, unglazed, and burnished. He further stated that with the exception of the pottery described by Binford, no Indian sites have produced Colono ware. Some Chesapeake archaeologists would take issue with this point, but it seems that their evidence, as far as we know it, is plain pottery being made by Indians in the region in historic times that resembles Colono ware but is not identical to it. So what we are left with in connecting this pottery with Indians is evidence that they made pottery in European shapes on the Gulf Coast in the early eighteenth century and burnished pottery similar to Colono ware in the nineteenth century. The tie between the pottery and Indians during the very years when it was manufactured in the greatest quantities is either "very weak," to quote Ferguson, or nonexistent.

This leaves the third group, African slaves, as possible candidates for the makers of Colono ware. Again, the evidence is circumstantial, but we can at least attempt a best-case scenario, one that achieves the closest fit with known historical and archaeological facts. Ferguson described Stanley South's excavations of the fortification ditch at Charleston, South Carolina, cut in 1670 and filled shortly after 1680. Colono ware was found in quantities in the fill of the ditch, telling us that this pottery appeared fully developed in the first decade of the colony. Yet in an adjacent Indian site, radiocarbon-dated to 1770 plus or minus eighty years, not a single sherd of this pottery was found; all of the ceramics were prehistoric in style. South Carolina had a very different pattern of population by Africans than did Virginia. Established at about the time that slave imports increased dramatically, it never witnessed a small black population as did Virginia during the earlier seventeenth century. Africans came into the colony in large numbers from the start, for most of South Carolina's first settlers came not from England but from Barbados, bringing their slaves with them. Slaves were making pottery in Barbados from at least as early as 1650, in both handmade and wheel-made forms. Pottery more similar to Colono ware than to plain Indian wares has been found in the West Indies on colonial sites. All of this evidence provides ample support for attributing South Carolinian Colono ware to Africans. Yet another aspect of the geographical distribution of Colono ware, not touched upon by Ferguson, is its absence from the northern colonies. There were Indians in the North, and also slaves. But the pattern of slavery there was a different one. Largely used for domestic service or as crew members on ocean-going ships, slaves were never housed in separate quarters in large numbers as was the case on southern plantations. Colono ware and the southern plantation economy seem linked if we consider their nearly identical distribution, and that link was probably provided by the common denominator of slavery.

Ferguson also told us that burnishing has a long tradition in African pottery making. In addition to burnishing, pots were decorated both by rouletting—rolling a serrated tool, perhaps like a watch cog, through the clay to produce dotted lines—and, after European contact, by rolling a corncob across the body of the pot. A corncob is a roulette of sorts, so the technique is but an extension of rouletting using a different tool. Maize was only introduced to Africa after European contact, and thus cob-marked pottery serves there as an important time marker, becoming very popular during the seventeenth century. This technique is also found on Indian pottery in the American colonies but only in the historic period, raising an interesting though somewhat speculative question. Of all pottery types made by Indians in the historic period in the entire Southeast, cob-impressed ceramics come from Virginia, the Carolinas, and Florida. These were areas settled early and centers of slave importation. Might it be possible that the technique was learned by Indians from runaway slaves? Colono ware, however, is predominantly undecorated, save for the burnished exterior surface. If it was being made for basic utilitarian use by the same people who would use it in most cases, decoration would seem hardly necessary. In Africa, on the other hand, pottery was produced in certain villages for distribution through trade over wide areas. In this context, decoration would make pottery a more desirable commodity and serve as an aid in the competition between potters of different villages.

Ferguson's second pair of questions, who used it and how was it selected, can be dealt with more briefly. He suggested that it was probably used by poor people, both black and white, and while this may be so, other evidence seems to rule out the possibility that poor Europeans used it. It is usually found in association with European ceramics of considerable quality, suggesting that its use was limited to the slaves of those who owned both them and the fancy European pottery with which it has been found. We badly need more excavations carried out on identified sites known to have been occupied by the poorest members of society. However, if Colono ware was made by slaves for their own use, it is unlikely that they would be producing a surplus to be used by poor Europeans. Which leads us to the question of how it was selected. Obviously, if it was made and used by slaves for their own needs, selection does not enter the picture.

So there the matter stands as far as the direct evidence is concerned. Proponents of the "Indians Made It" school base their arguments largely on the occurrence of plain pottery on historic-period Indian sites, similar but not identical to Colono ware and with considerable emphasis still on the Catawba and Pamunky material. Against this stands a compelling set of correlations, between the concurrent increase in Colono ware and the black population, between sites that produce Colono ware and an African presence on them, and between the geographical area of plantation slavery and that where Colono ware occurs. It is commonly found on European sites with an African presence and, with the possible exception of the material discussed by Binford, has yet to be found on Indian sites of the period. The identification of Colono ware as Indian-made is at least in part an accident of history. Had Noël Hume not been the first to call attention to this pottery and identify it as Indian, someone else might have arrived at the opposite conclusion. If that conclusion had been published, establishing the pottery as being made by slaves, would the counterargument have been put forth as vigorously as it has? The South Carolina evidence is stronger and more convincing than that from Virginia, though the evidence from the Chesapeake seems strong enough to stand on its own. But had it been initially identified as a slave product, would the development of the pro-Indian argument have taken the same course? We will never know, although it seems unlikely and makes for interesting speculation. But the case has not yet been presented fully, and we must now return to Flowerdew Hundred to develop it further.

The eighteen sites scattered across Flowerdew Hundred's bottomland provide us with an unusual measure of control, since they represent a continuous occupation from 1619 through most of the eighteenth century, and even include one nineteenth-century site. This permits us to observe change and variation in a single location and not be concerned that any difference between house types, artifacts, or settlement layout is

the result of the things compared being in different locations. The archaeological dimension of space is thus tightly controlled, allowing us to explain differences only in terms of time.

With this in mind, we recall that the five sites in the third group all
show a rapid development during the opening years of the eighteenth
century. It has been suggested that this is a function of the great increase
in the number of slaves arriving in Virginia, and that the occupants of
these sites at that time were modest slaveholders. If Colono ware was
made and used by slaves, this conclusion is correct, for all five sites produced that pottery while none of the earlier sites did. Even those who
maintain that Indians made Colono ware agree for the most part that
slaves were its primary consumers, and so a slave presence is still indicated
on the sites from the early eighteenth century onwards. Documents also
tell us that slaves were at Flowerdew at this time, for they are mentioned
in wills and probate inventories. The evidence is less clear in the years
before 1680, but in light of the fact that numbers of Africans were at
Flowerdew Hundred almost from the time it was established, the likelihood of a continuous black presence on the plantation is high indeed. If
this is the case then, we must ask why no Colono ware has been found
on Flowerdew sites that predate the final two decades of the seventeenth
century. To answer this question, we must return to the thesis developed
by Dell Upton concerning servant-master relationships.[8] The full establishment of slavery as an institution based solidly and solely on race did
not take place in Virginia until after 1680. Before that time, attitudes toward blacks, whether slaves or not, were more flexible and variable. Edmund Morgan has explained that "while racial feelings undoubtedly
affected the position of Negroes, there is more than a little evidence that
Virginians during these years were ready to think of Negroes as members
or potential members of the community on the same terms as other men
and to demand of them the same standards of behavior. Black men and
white serving the same master worked, ate and slept together, and together shared in escapades, escapes and punishments."[9]

Morgan pointed out that before 1660 some of the Negroes, perhaps a
majority, were slaves, but others were servants and some were free. They
could earn wages and even use the money to purchase their freedom. All
of this changed later in the century, and by 1700 the vast majority of
Africans in Virginia were enslaved because of their color. Given such a set
of social relationships between black and white before 1680, it is not at
all unreasonable to suggest that Africans were housed under the same roof

as their master, for Upton has shown that separate quarters for servants and slaves do not make their appearance until the last two decades of the century. What has this to do with Colono ware? If a slave or African servant was living in his or her master's house, there would be little or no need to produce pottery in which to cook and from which to take one's meals. The pottery used would be furnished by the master and would be of European origin. Furthermore, blacks in such a situation would be far more likely to learn European ways of preparing and serving food and become very familiar with porringers, pipkins, flat-bottomed serving bowls, and the like. But upon being settled away from the main house, all that would have changed, because except for a roof over their heads and perhaps an ax and iron pot, slaves were given little in the way of material objects. In these new circumstances, slaves would need to produce their own pottery, in a tradition with which they were already quite familiar.

All of which leads us to the differences in shape. Leland Ferguson found that the careful replication of European shapes seen in Virginia Colono pottery is almost completely lacking in the South Carolina version of the ware. With a few exceptions, three basic shapes were produced in South Carolina: large globular pots, small globular pots, and shallow bowls. The pots, both large and small, show evidence of use as cooking vessels. The bowls show considerable abrasion on their rims, possibly made by utensils used in consuming the food they contained. This trio of shapes fits exactly with the basic food preparation and consumption assemblage found over all of West Africa. The large pots are used to prepare whatever starchy food that might be used, such as manioc, samp, or rice. A mix of vegetables, meat or fish, and spices is cooked in a smaller pot, and the dish is served by placing the starchy base in a shallow bowl and spooning the meat and vegetable sauce over it, consuming it with the fingers.[10] This way of cooking and serving food is found in modern American cuisine in the form of gumbo, a dish widely regarded as having African roots; it uses okra, a plant introduced from Africa, as a thickener. This manner of food production and consumption, as well as the pottery used, is just what one would predict given the circumstances of early black settlement in South Carolina. Unlike Virginia, where small numbers of blacks were present throughout the seventeenth century, Africans arrived in South Carolina late and in great numbers from the start. Settled at the outset in quarters at some distance from the master's house, they would never have had the opportunity to become familiar with European pottery forms and the dishes prepared in them and consumed from them. South Carolina and Georgia are areas where there has been a strong retention of a variety of other African cultural elements as well. These include the coiled basketry made on the Sea Islands; the Gullah language, a creole using a modified English vocabulary set in a grammatical context more West African in form; a strong woodworking tradition; and African methods of grave decoration.[11]

So it is that the differences in shape of Colono ware from the Chesapeake and the Carolinas can be best explained by the differences in the way blacks and whites interacted in the two areas. This relationship in turn strengthens the identification of Colono ware as having been made by slaves, for it is difficult to account for the differences that we see if Indians are cast in this role. And perhaps most important of all, it lends archaeological support to the evidence that Morgan cited as indicating the nature of black-white social interaction in the years before 1680. The historical record for this period is not as strong as we would like on the subject, so that any shred of information is valuable indeed. It forces us to reexamine our ideas about slavery and black servitude in the years before it became cast in the form that we are most familiar with, that of the antebellum plantation made famous by Margaret Mitchell and Alex Haley, a picture that we project on the more remote past in an uncritical fashion. Slavery as it was known in later times did not arrive full-blown on the Virginia shore in 1607. The first blacks to make their appearance arrived in 1619, and we have met some of them at Abraham Peirsey's Hundred. It is generally believed that these first Africans were not slaves but servants. They had Christian names, and at that early date religious faith was a far more significant criterion than color of who could be enslaved and who might not, "heathens" being excellent candidates over fellow Christians. This attitude changed as more and more Africans arrived in the Chesapeake, so that race rather than belief emerged as the primary criterion for defining who was a slave and who was not as the eighteenth century entered its first years. Other Africans followed these first black Virginians, but not in great numbers, perhaps constituting a population of less than a thousand by the mid-seventeenth century. The first six or seven decades of that century was a time of experimentation, of the slow but steady emergence of slavery as a formal institution, and for that reason also a time when black-white social interactions were more tentative and fluid than at any time since. And as we have seen, archaeology does shed some light on the nature of these relationships during these very critical years. Yet another dimension is provided by the pipes from the cellar of site 77, for their maker's identity is also at issue in modern archaeological discourse.

The ubiquitous white clay English smoking pipes played a key role in the development of the Flowerdew Hundred story. But beginning sometime in the 1640s—the precise date cannot be determined—another kind of pipe made its appearance in the Chesapeake. Made from local tidewater clays, these pipes are easily distinguished from their English counterparts (fig. 17). They are a red-brown color, sometimes referred to as terra-cotta, and have thicker stems than the English ones, perhaps because a stiffer, less malleable clay was used in their manufacture. The bores of the stems are also larger, and as a result such pipe stems cannot be used for dating purposes. Some seem to have been made by hand, but the majority of them appear to have been made in a mold, following the English method of production. Shapes include those identical to English pipes, as well as others that are somewhat different but still resemble English examples in a general way. What sets them apart from English pipes most is the manner in which they were decorated, with elaborate geometric and naturalistic designs executed by both stamping and rouletting on bowls and stems. The rouletting was done with some implement with very fine teeth, which produced lines of closely and evenly spaced small punctates, which were then filled with a white substance, either clay or lime. This white infilling is quite fragile and has disappeared from many pipes, but chances are good that almost all such decorated pipes originally had white infill in the rouletted design. The finished product is quite attractive, with the white design standing in pleasant contrast to the tan body of the pipe. Fish, horned animals, stars, ships, and a variety of geometric motifs appear on the pipe bowls, and the stems are usually decorated in simple designs of lines and triangles. How do these pipes relate to the development of local industries, and who was responsible for their production?

In answering this question we are drawn into a difference of opinion which parallels that concerning the makers of Colono ware, and indeed the two problems can be shown to be closely related, forming parts of a larger whole. Unlike Colono ware, these pipes are found only in the Chesapeake, east of the fall line. Within that area, any site that dates to their period of production, roughly between 1640 to 1720 at the latest, will produce fragments of them in some quantity. The largest number of them have come from Jamestown, where they made up 22 percent of the total number of smoking pipes, and they can occur in amounts as low

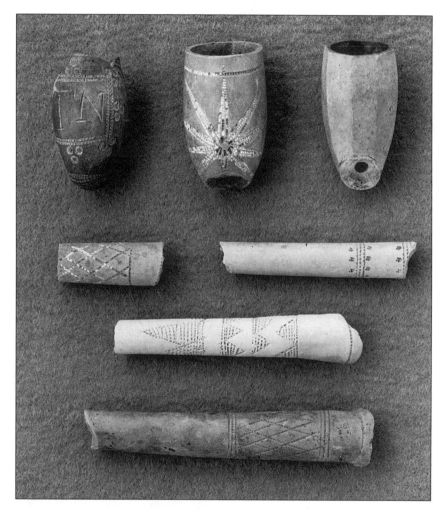

17. Chesapeake pipes from Flowerdew Hundred, with characteristic decorations on bowls and stem fragments

as 1 percent, with the average quantity being of the order of 6 percent. Significantly, they are not found on early sites with no later occupation, such as Martin's Hundred's Wolstenholme Town or the enclosed compound at Flowerdew Hundred. The cottage industry that these pipes represent is but one of several that arose at that time in the Chesapeake, including bloomery iron production such as that evidenced at Flowerdew Hundred. Thus they were yet another attempt at self-sufficiency during the second half of the seventeenth century. They seem to have been produced at a relatively small number of locations, although this is largely an intuitive judgment; careful analysis of the distribution of various design types would go far to clarify this supposition. Like Colono ware, they had been found on sites throughout the region for years, but until 1979 little more was said about them than noting their presence and describing them. They are often referred to as Colono-Indian pipes or Colono pipes, terminology directly related to that used for the pottery.

As in the case of Colono ware, when the first extended treatment of these pipes appeared in a 1979 article by Susan Henry entitled "Terra-Cotta Tobacco Pipes in 17th Century Maryland and Virginia," they were identified as having been made by Indians.[12] Henry's main question was not who might have made the pipes, but whether their production was related to economic changes. She proposed that in times of shortages of imports, which would have resulted from a drop in tobacco prices, production of these pipes would increase, and she found such a correlation. But she handled the question of the identity of their makers in a way reminiscent of Noël Hume's attribution of Colono ware to Indians and indeed based her argument on his earlier work: "In view of the fact that they [Indians] made ceramics in imitation of European forms (Noël Hume, 1962:3), they probably also made tobacco pipes resembling European forms which might have been sold or traded to the colonists."[13] We can see then, that the identification of these decorated pipes as Indian made is but an offshoot of Noël Hume's identification of Colono ware, and thus is also connected back to Speck's Pamunky work, published in 1928. But a close examination of Henry's study reveals some interesting patterns of correlation that point in quite a different direction.

Henry's study is based on the analysis of 111 specimens recovered from the site of Saint John's, at Saint Mary's City, Maryland. She suggested that two traditions of local pipe production coexisted in the Chesapeake, one of pipes made in molds by English pipe makers and the other of handmade pipes, produced by Indians. The pipes that she identified as handmade are decorated with rouletted designs and have the characteristic white infill, and those identified as mold made show stamped designs, typical of English pipes. She developed a classification of pipe shapes and decorative motifs which is presented in tabular form (table 1), with drawings of shapes and designs (fig. 18). According to this classification, bowl shapes C through F are identified as those of handmade pipes, and shapes, G, H, and I are of mold-made ones. But this table can be read in a very

Table 1. Distribution of design by bowl shape in terra-cotta pipes from St. John's, St. Mary's City, Maryland

Design	A	B	C	D	E	F	G	H	I	Total
*	–	–	–	–	–	–	–	–	–	0
2	–	6	2	4	2	–	–	–	–	14
3	–	–	4	1	–	–	–	–	–	5
4	–	–	11	2	–	–	–	–	–	3
5	–	–	6	8	–	–	–	–	–	14
6	–	–	5	1	–	–	–	–	–	6
7	1	–	1	1	–	–	–	–	–	3
8**	–	–	10	1	7	2	1	9	5	35
9	–	–	–	–	–	–	–	6	–	6
10	–	–	–	–	–	–	1	–	–	1
11	–	–	–	–	–	–	–	8	3	11
12	–	–	–	–	–	–	–	3	–	3
Total	1	6	39	18	9	2	2	26	8	111

Source: Reprinted by permission of the Society for Historical Archaeology from Susan L. Henry, "Terra-Cotta Tobacco Pipes in 17th Century Maryland and Virginia: A Preliminary Study," *Historical Archaeology* 13 (1979): 14–37.
*Design motif 1 (punctate) occurs only on a few stem fragments.
**No decorative motif—i.e., plain.

different way, one that better fits certain other aspects of both the archaeological and historical record. At the outset, the suggested opposition between handmade and mold-made pipes each with their own set of designs is not reflected in the hundreds of decorated bowls recovered from Flowerdew Hundred sites. It is often very difficult to tell whether a pipe bowl came from a mold, but one bit of telltale evidence shown by the Flowerdew pieces is a little scar inside the bowl, opposite the point where the stem bore passes through. This scar is the result of the wire passing through the stem while in the mold and pricking the clay at the inside base of the bowl. Handmade examples do not show this feature. While burnishing often erased any lines on the exterior that resulted from the two-piece mold, this bit of evidence remains inside. Large numbers of rouletted bowls similar to Henry's shapes C through F show these scars, indicating that pipes of these shapes were made using molds. This evidence leads us to suggest that mold-made pipes were being produced during the entire period when rouletted designs were being applied to their bowls.

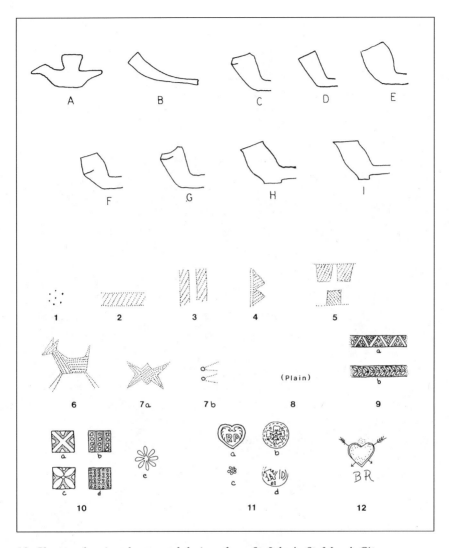

18. Chesapeake pipe shapes and designs from St. John's, St. Mary's City, Maryland. (Reprinted with permission of the Society for Historical Archaeology from Susan L. Henry, "Terra-Cotta Tobacco Pipes in 17th Century Maryland and Virginia: A Preliminary Study," *Historical Archaeology* 13 [1979]: 14–37)

From Henry's article and the accompanying tabulations, it is possible to suggest a three-phase sequence exhibited by these pipes. The earliest is quite different from the following two and represents pipes made by local Indians. These are represented by bowl shapes A and B. Shape A is represented by a single example; it is a bird effigy platform pipe, a common Indian form. Bowl shape B, represented by six fragments, is Indian made as well, a typical late Woodland tubular pipe. The pieces were recovered from a 1640–60 context at Saint John's and are probably intrusive, since it is common for native artifacts to become included in fill deposited later, in the colonial period. But even if the association is valid, and it could be since Indians were present at the time, its distinctive and unmistakably Woodland shape sets it apart from bowl shapes C through L, all of which are either very similar or identical to pipes of English manufacture.

The other two phases are represented by the remaining eight bowl shapes, and except for design 8, which is not a design but rather indicates a lack of decoration, designs 1–7 and bowl shapes C–E form a set that has no overlap with the set formed by designs 9–12 and bowl shapes F–I. The correlation is perfect. Of importance here is the fact stated by Henry that pipes of the first group seem not to have been made after 1680, while those in the second group continued to be produced into the first two decades of the eighteenth century. If these pipes were Indian made, why does their production cease when it does? And is it a coincidence that Colono ware begins to appear just when these rouletted pipes vanish from the archaeological record? The first step in providing an answer is to consider, if only for sake of argument, that someone other than Indians was producing them in great numbers.

The first person to suggest this possibility was Matthew Emerson in a 1988 University of California, Berkeley, doctoral dissertation entitled "Decorated Clay Tobacco Pipes from the Chesapeake."[14] In this work he proposed the adjective *Chesapeake* as a replacement for the older terms *Colono-Indian* or *Colono*, an appropriate designation because it does not imply an ethnic identity for the makers of the pipes, and it also frees them from the modifier *Colono*. Emerson was the archaeologist who excavated site 77 at Flowerdew Hundred, and his interest was aroused by the large numbers of Chesapeake pipes that came from the fill of the enigmatic cellar. He was the first to attempt an exhaustive survey of all of the accessible collections in the region and also the first to compile a complete corpus of the designs found on the pipes, in terms of motif and technique of production. Armed with this impressive mass of material, he then set about to examine all of the designs in an effort to find correspondences with English, African, and native American elements. Two sets of designs eventually emerged from this survey and classification. The first consisted of designs that occur in decorative traditions in many parts of the world, including both West Africa and North America. These include "hanging triangles," a band of triangles suspended apex down from a line, and parallel lines, both common in Nigerian and Algonquian decorative arts and not sufficiently complex to point to any specific tradition. The second group of motifs, however, are designs of considerable complexity, unlikely to result from coincidence, and not part of the local Algonquian design vocabulary. For these latter motifs Emerson found parallels in the decorative arts of West Africa, leading him to suggest an African component in the designs placed on Chesapeake pipes. While these designs do not constitute exclusive proof that Africans decorated the pipes, they at least cause us to look at the problem from a new perspective.

The most convincing of the motifs is the so-called Kwardata, a highly individualistic design which is identical in Chesapeake and West African examples. In Nigeria the Kwardata design symbolizes the transition from youth to adulthood. The motif is created by defining a connected band of diamonds with parallel punctate lines which act as a background, the design itself being negatively formed (fig. 19). This method of motif definition is not a part of the local Indian tradition, and it is so distinctive that its occurrence in identical form in both Chesapeake pipes and Nigerian decoration would seem to be the result of a genuine relationship between the two. Equally striking is a design formed by several concentric arcs, with a row of circles above them. Like the Kwardata motif, this design is highly individualistic and lacking from the Indian repertoire but has an African parallel (fig. 19). The "double bell" motif, two arcs flanking a horizontal line, is found both on Chesapeake pipes and on a smoking-pipe bowl from Cameroon (fig. 19).

Less convincing but suggestive nonetheless are two other designs, stars with sets of circles at each of their points and "quadrupeds," depictions of some kind of horned animal. The stars are not a part of the Indian tradition of ceramic decoration, and the use of circles to embellish end points of various motifs is common in African decorative arts. However, stars have a wider pattern of use worldwide and are not as idiosyncratic as the first three designs. The quadrupeds are another matter; while depictions of four-legged animals probably occur in any naturalistic design vocabulary, the horns on many of the animals found on Chesapeake pipes

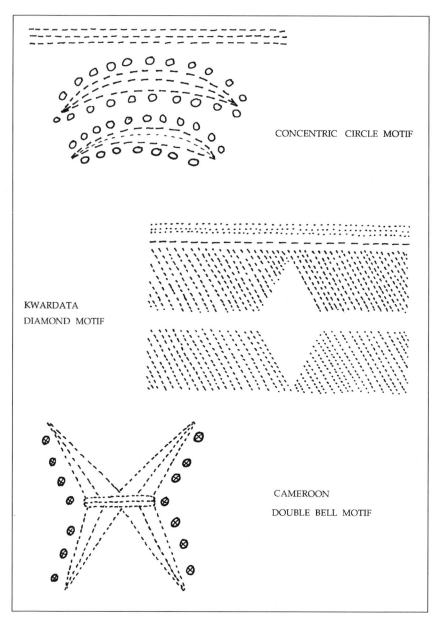

CONCENTRIC CIRCLE MOTIF

KWARDATA
DIAMOND MOTIF

CAMEROON

DOUBLE BELL MOTIF

19. African motifs seen on Chesapeake pipe bowls: concentric arcs
and dots; Kwardata motif; "double bell"

suggest a non-Indian derivation. These animals have been identified by archaeologists as deer, and in fact one name given to the motif is the "running deer" design. Such an identification is logical if one assumes that Indians executed the design, for deer would be the only horned animals that they would have known. But the animals on the pipes have horns that look far more like those of some kind of antelope, or at the very least a goat, and neither of these animals were known to the Indians before European contact. Goats would have been seen by Indians during historic times, but it seems likely that the designs were executed by people who had a long association with goats, if that is what they are meant to be, or an exclusive knowledge of antelopes, if they are the animals depicted. In either case an African derivation of the quadruped motif is not at all unreasonable, and the other, more specific resemblances between Chesapeake pipe designs and African examples lends strength to such a suggestion.

As convincing as these resemblances might appear, alone they are not sufficient evidence to permit the attribution of Chesapeake pipe decoration to African artists. What they do provide us with is a basis on which to explore further the possible relationships between these decorated pipes and blacks in seventeenth-century Virginia and Maryland. The questions posed earlier regarding the relationship, if any, between the pipes and Colono ware can be considered in the light of both identifications, Indian and black, to see which best fits the evidence at hand.

It is virtually certain that Chesapeake pipes were manufactured for sale. This is not only evidenced by Henry's correlation between numbers of pipes and fluctuations in the price of tobacco but also by the fact that unlike Colono ware, they required a kind of technological sophistication probably not possessed by all members of Chesapeake society. While the production of pipes using molds is not complex, it is more involved than the simple construction of handmade pots. Certain basic equipment is needed, as well as the knowledge of how to use it. Pipes of this sort can be turned out in great numbers in a short period of time by only a few people. It is for these reasons that one gets the sense that they were made in a few locations and distributed widely over the region. They rarely constitute the majority of pipes on a site, and it could well be that they were produced to make up a shortfall in imports at certain times. Clearly large numbers of them produced using a technology known only to the colonists, and so whoever was applying the decoration either had learned the method of making them or was decorating pipes made by

English pipe makers. In these aspects the context of production is quite
different from that of Colono ware, which involves no European technology whatsoever. They are a kind of hybrid in a way, with the method of
manufacture being different from the means of decoration and the designs used. Rouletted designs such as those found on Chesapeake pipes
have never appeared on pipes of English manufacture. Thus, these motifs
could be of either African or Indian origin, or perhaps both groups were
involved in their execution.

But beyond the design elements shared by the pipes and West African
decorative arts, the manner in which the pipes and Colono ware appear
over time allows us to put forth an argument for African decoration on
the pipes which better fits the facts at hand. At Flowerdew Hundred we
have not yet found sites which produce both Colono ware and Chesapeake pipes. Such sites do exist elsewhere in the Chesapeake, and any
site occupied continuously between the mid-seventeenth century and the
beginning of the eighteenth could be expected to yield both types of artifacts. Yet sites predating ca. 1680 will produce rouletted Chesapeake pipes
but no Colono ware, and conversely sites dating after the 1680s and occupied into the eighteenth century will show the opposite pattern. What
stands out clearly in this complementary distribution over time is that
the pipes and the pottery appear to represent the same thing, in a sense:
artifacts partly or wholly non-European in origin but produced in the
context of colonial society for the use of members of the plantation community, either European or African.

If Indians were responsible for producing Colono ware and decorating
Chesapeake pipes, it seems odd that they would turn from pipe decoration to pottery manufacture as the seventeenth century came to a close.
One could argue that Indian servants or slaves worked with English pipe
makers before that time, but we must remember that those who argue for
Indian production of Colono ware suggest that it was not made on the
plantations but in villages situated at a considerable distance from them.
This forces us to ask why Indians waited until the 1680s to make Colono
ware, if they had been making pipes since the 1640s in "forms which
might have been sold or traded to the colonists."[15] This question becomes
even more relevant when we recall that local production of pipes continued into the eighteenth century (Henry's bowl shapes G–I and design
types 9–12), while the classic roulette-decorated pipes in non-European
motifs fade from the picture ca. 1680.

A far more economical and convincing explanation of the pattern manifested by these two types of artifacts can be set forth by attributing both the pipe decoration and pottery manufacture to Africans. Recalling Dell Upton's study of house size and master-servant/slave relationships, it makes sense to suggest that during the time that blacks and whites were residing together, they would also have had a far better opportunity to engage in craft production that would combine elements of both cultures than would have obtained after the breakdown of what had once been a closely integrated community. The social context that would have fostered the production of pipes made in a European technology and decorated with African designs was replaced by one that would have brought this to an end, to be replaced by a set of relationships in which the production of Colono ware would have become almost a necessity. Only this scenario accommodates all of the various pieces of the puzzle in a convincing manner, freeing us from suggesting that in some complex and obscure way, Indians either on the plantations or in their own villages somehow agreed to stop making pipes and begin the production of pottery for some mysterious reason during the 1680s. Africans on the other hand are the common link between the two types of artifacts, and given what we know of the changes which were worked on their pattern of residence in the late seventeenth century, the explanation offered here has considerable merit.

Finally, we must note that the geographical distribution of Chesapeake pipes is exactly that of the earliest black presence in colonial America. During their period of production, only the area in which they have been found had a population, albeit small, of Africans. No pipes of this type have been found either in the northern colonies or to the south in the Carolinas and Georgia, areas with an equally significant Indian population at the time. There are locally made pipes from New England sites, but they are quite different, undecorated and made from a distinctive red clay. They were produced during the same time period as were their Chesapeake counterparts and probably for the same reason, periodic shortages of imports from England.

Taken together then, the evidence which we presently have indicates African authorship of both Colono ware and Chesapeake pipe designs. If so, the designs are among the earliest examples not only of American folk art but of African American folk art in particular. The argument will continue to be mounted by proponents of both points of view, and while that set forth here may or may not be totally convincing, it has a certain logic not to be denied. But until a piece of critical evidence is discovered, perhaps a kiln site where Colono ware was produced or a massive dump of pipe wasters on a plantation site, with both broken molds and rouletting tools, debate over the question will continue. This is as it should be, for the implications of the questions as they relate to the larger social context of the Chesapeake are important, and archaeology will continue to play a central role in their resolution.

By the third quarter of the eighteenth century, the Flowerdew Hundred bottomlands were largely unoccupied and probably were continuous fields of tobacco, and possibly corn and wheat as well, looking much as they do today. The last remaining settlement was at the far southern end of the plantation, and it is here we must now look, for one of the sites there may hold a clue as to what happened to Powhatantown. Three sites figure in this account, numbers 97, 98, and 103. Situated quite close to one another, they represent a set of structures so closely related in time and function as to warrant designating the entire group by a single number, which was done by Flowerdew archaeologists in the summer of 1991. Site numbers 97 and 103 will be retained for the purpose of recording earlier collections from those sites, but henceforth all data will be subsumed under the single site number 98, and that number is used in the discussion here. Site 98 is by far both the largest and the longest-occupied at Flowerdew Hundred. There are faint traces of a pre-1650 presence in the materials collected from the surface of the site, the odd bit of early pottery and occasional pipe stems with bores of $\frac{9}{64}$ and $\frac{8}{64}$ inch. These pieces could simply be spillover from site 86, one of the seven early sites, located not far away on the river shore. More common are artifacts from the second half of the century, including Chesapeake pipes and typical late seventeenth-century ceramics. But the vast majority of artifacts collected and those structures thus far excavated show an occupation between the beginning of the eighteenth century and the first decade or so of the nineteenth.

It appears that such a prolonged presence in one location resulted from something other than tobacco cultivation or experimentation with some manufacturing activity. Fortunately, the records are most illuminating in this respect, showing that a ferry service was operating at this location throughout the eighteenth century. The name of the ferry fits the known facts of property ownership on the southern portion of the plantation. Joseph Wilkins acquired land there from his father John Limbrey Wilkins, a member of the Limbrey family who held title to the southern half of the property from John Barker. There are only two detailed descriptions of eighteenth-century structures at Flowerdew Hundred, and the Wilkins dwelling house is one of them. (The second is the house built by William Poythress on the ridge to the west, and last occupied by Miles Selden.) Wilkins died by 1767, and on September 24 a notice appeared in the *Virginia Gazette*:

> To be rented to the highest bidder, on Saturday the 26th of this instant, the plantation, containing 138 acres of exceeding good land and Ferry at Flower de Hundred, late belonging to JOSEPH WILKINS, deceased, wheron is a good dwelling house, with three rooms on a floor above and below, and a good brick cellar underneath; likewise a good new kitchen with a brick chimney, and a good apple orchard. The ferry is about three quarters of a mile wide, very convenient, and the nearest way from Petersburg to Williamsburg; and if well kept, will raise 30 or 40 l a year. The plantation is exceedingly good for raising cattle and hogs, on account of its nearness to Flower de Hundred marshes, where there is an excellent place for sturgeon fishing; as between 60 and 70 have been caught at a haul. Bond and security given to
>
> George Noble   Administrators
> 'Joshua Poythress

Exactly five years later, on September 24, 1772, Sherwood Lightfoot placed an announcement of his own in the *Gazette:* "The subscriber having rented Wilk's ferry and ordinary at Flower de Hundred, takes this opportunity to inform the public that he will be there by the October General Court, where such Gentlemen as please to favor him with their custom may depend on good accommodation for themselves and their horses, and a ready passage. He flatters himself that he shall give entire satisfaction, which may induce many to cross at his ferry, as it is certainly, by six or seven miles, the nighest way from Petersburg to Williamsburg." Two years later, John Crosby told the public that he had leased the property: "The subscriber has lately opened Tavern at Wilkin's Ferry, where all Gentlemen Travelers to please to favor him with their custom may depend on good accommodations and a quick passage. . . . John Crosby." [16] Other lessees of the ordinary and ferry followed, and the estate administrators George Noble and Joshua Poythress continued to advertise from time to time in the *Virginia Gazette*, the last time in 1779.

It is quite clear from the descriptions of the property and the activities that were taking place there that it had become a service facility, a way of making an income without investing in cultivation or manufacturing. It is this that accounts for the site's long occupation, for as long as there were people traveling from Williamsburg to Petersburg and back, there would be a need for such accommodations as well as transport across the river. The 1767 notice says that the crossing is about three quarters of a mile wide, which is the distance across this portion of the James River today. It is much wider both below and above this point, a fact that was to become very significant nearly a century later when General Grant crossed the James just south of site 98.

Surface collections made in the 1970s lend further support to the existence of a tavern on site 98. Most sites that once held dwelling houses will produce pipe stem fragments numbering in the low hundreds at most, and some only fifty or so. But site 98 has produced more than two thousand stem pieces from the surface alone. Fragments of German stoneware mugs are also quite common, as well as wine glass bases of the so-called tavern type, which are thick and heavy. So we have excellent evidence both from the documents and the artifacts that a tavern once stood somewhere on site 98, as well as other structures that would have supported it. But fitting what the archaeology has shown with the documents is not quite as straightforward as one might like, although the general agreement is good, and there is no question that the site and the property mentioned in the newspaper notices are one and the same. Excavation at site 98 is by no means complete, but thus far the remains of four structures have been unearthed. They answer some questions and raise new ones as well. These buildings align themselves along a north-to-south axis, with the earliest one situated at the southern end, the second two oldest in the middle, and the latest at the northern end. Thus the site shows a general south-to-north pattern of growth. Three incorporate brick construction of some kind or another, and their existence had been known for years because the plow had been striking the brickwork just below the surface. The fourth, southernmost, structure is an earthfast building.

The first site excavated was the northernmost one (formerly site 97); it proved to be the foundation of a one-room house with a chimney on the northern end and a full brick cellar entered by a set of stairs through a bulkhead entrance at the southern end. The house had burned in a very hot fire sometime in the first twenty years of the nineteenth century. The fire was so hot that it stained the ground a bright red to the depth of subsoil. No effort seems to have been made to retrieve any of the nails or building hardware from the ashes, and over twenty thousand nails of various types were recovered as well as a variety of hardware, such as hinges, pintles, and lock plates. These iron artifacts look as if they were made yesterday; the hinges still swing freely, and the nails are totally free of rust. Hundreds of mud wasp nests were also found in the cellar, fired to the hardness of pottery. Since the farmers had been lifting their plows while passing over the foundation, the area just outside the bulkhead entrance showed an unusual feature, the faint traces of colonial plow scars running at right angles to the direction taken by modern plows. These plow scars predate the construction of the house, since the bulkhead entrance cuts through them. Ceramic evidence as well as pipe stems suggest an initial occupation of this house sometime in the late 1760s or 1770s, with Staffordshire white salt-glazed stoneware providing the best *terminus post quem*. The latest pottery recovered is that known as pearlware, in types typical of the first decade or so of the nineteenth century. The identity of this building is less than clear from the documents. It could have been standing at the time of the 1767 notice, but a much more likely candidate for the dwelling house described in that advertisement is the third structure to the south. For this reason the excavators concluded that it was the "good new kitchen" which the notice also mentions. Certainly the dating evidence indicates that it was new in 1767; in fact, it may not even have been built that early. Moreover, most of the pottery fragments recovered from the cellar and surrounding area were from plates and cups, serving pieces rather than what a kitchen might produce, such as cooking vessels or pottery for food storage. On balance, it seems more likely that the building was a dwelling house, leaving the location of the new kitchen yet to be discovered.

South of this house is a second foundation, with a massive chimney on the southern end. The foundation is quite small; when excavated it was found to measure twenty feet in length and thirteen feet in width. Save for the chimney base and small bits of the cellar brickwork tied into it, the brick that had once lined the cellar of this structure had been removed, probably to be used in construction elsewhere. The open cellar was used as a refuse dump and contained a dense mass of artifacts dating to the last quarter of the eighteenth century. Other artifacts, recovered from contexts that represent the occupation of the building, such as those from the chimney base fill, suggest a date somewhere between ca. 1750 and 1800 for the structure. Far more than the house just to the north, this small building with its large chimney could well have been a kitchen, but the question remains as to which house it was associated with. Both structures are oriented in identical fashion, almost on a due north-south alignment. In this, they differ from the third and fourth structures to the south, which are angled some ten or so degrees off the north-south line.

The third structure is represented by a brick cellar, measuring sixteen by twenty feet at a depth of two and a half feet below subsoil. The fill of this cellar was deposited in three layers. The lowermost consisted of artifact-rich fill, including ceramic types typical of the second and early third quarter of the eighteenth century. Overlying this layer was a thick deposit of bricks from the upper walls of the cellar, pushed in after the building was abandoned. The top layer was a fine sandy deposit, poor in artifacts but producing examples of creamware that allowed us to date the final filling of the open cellar sometime during the 1770s. The cellar floor had been purposely built up with a four-inch layer of packed clay, presumably to accommodate heavy use, since it capped an underlying deposit of undisturbed clay with a high sand content, the natural base of the hole excavated for the cellar. As in the northernmost building, the cellar entrance was through a bulkhead entrance from the south. However, this entrance was significantly wider, possibly to allow large hogsheads to be rolled in. On the floor was a three-pointed cooper's tool, used to cut bungholes in barrels. No evidence of a chimney has survived, although a heavier concentration of mortar and brick fragments was encountered in the plow zone just off the northern end of the cellar, and two bricks were bonded into the cellar walls, projecting to the north. The cellar is probably missing at least ten courses of brick around its top; had the chimney been erected around a hearth built on the original surface at the level of the top of the cellar, evidence of it would have been removed by centuries of plowing. The two brick cellars are almost identical in form, and one of them must be the remains of the ordinary advertised in the *Virginia Gazette.* But before we consider this problem, we must account for the fourth and earliest building of the four.

Immediately adjacent to the southernmost brick cellar and extending at right angles from it to the east, a thirty-by-sixteen-foot earthfast structure once stood (fig. 20). It is oriented identically to the brick cellar, but ten degrees off the northernmost pair of cellars, so that each pair forms a separate set. This house was built of large posts set on ten-foot centers. The hearth base was found on the eastern end, as well as postholes indicating a wooden wattle-and-daub chimney. A shallow cellar abutted the hearth, and a second, deeper cellar was located on the western end, accessible through a bulkhead entrance. This second cellar had a floor of wooden planks set on joists which in turn rested on the earth floor. Adja-

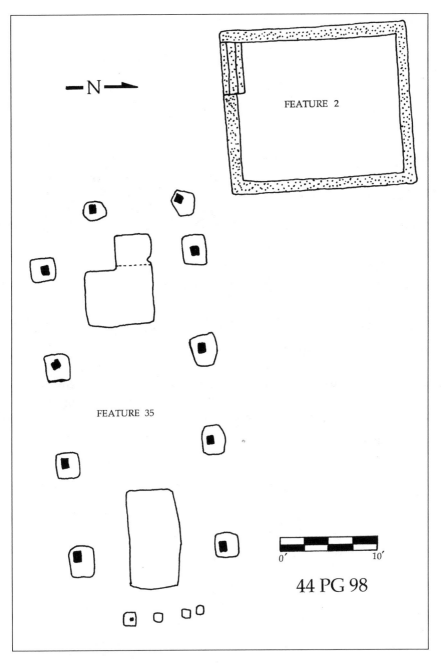

FEATURE 2

FEATURE 35

44 PG 98

0′   10′

20. Earthfast building (feature 35) and brick cellar (feature 2), site 98. Feature 2: the brick cellar is that of the Wilkins's Ferry ordinary.

cent postholes indicate a shed addition projecting to the west, covering the cellar entrance. This building appears to have been constructed at the turn of the eighteenth century and stood until sometime in the 1740s. A Spanish real with the date of 1736 was recovered from the fill of the eastern cellar, and pipe stems and bowls as well as ceramics support a date of the first half of the eighteenth century, and probable destruction before 1750. Since the two structures represented by the brick cellar and the hole-set house would almost have touched if standing at the same time, and since the artifacts from the fill of each indicate sequential building, it seems reasonable to suggest that when the earthfast dwelling house was torn down, the house represented by the brick cellar was erected. The ninety-degree relationship between the two structures is also relevant, since earlier houses frequently faced south, while later ones faced a road or, in this case, a road's equivalent, the James River just to the east. The house which is represented by the postholes was a rather well constructed one; pieces of wall plaster and numerous window leads were found in the eastern cellar fill. An adjacent trash pit, which also contained mortared bricks, possibly from the hearth, and plaster fragments as well, produced a piece of window lead which when unfolded showed the initials EW and RA on the inside, accompanied by the date 1693.

Such dated pieces of window lead have come to light recently on a number of Chesapeake sites and have been found on English sites.[17] They provide us with a fascinating look into problems of quality control in an emerging industry in seventeenth-century England. This window lead is in the form of strips, H-shaped in cross section, known as cames. Windows of the period had small diamond-shaped panes held together by lead, with the edges of the panes resting in the grooves on either side of the cames (fig. 21). It is on the inside of the grooves in the cames that we find initials or names and dates. One wonders why came makers would place their name or initials as well as a date in a place where it would not have been seen unless the came were removed and carefully unfolded to reveal the inscription. The reason seems to have something to do with problems of substandard cames and manufacturer's liability.

Although such inscriptions have been known since 1912, interest in them was renewed when they were discovered on archaeological sites in the Chesapeake. Ivor Noël Hume discovered a very early example at Martin's Hundred which read, *"Iohn: Byshopp of Exceter Gonner: 1625."*[18] Later ones have been found at Saint Mary's City, Maryland. It is estimated that approximately 10 percent of lead cames at American sites bore such in-

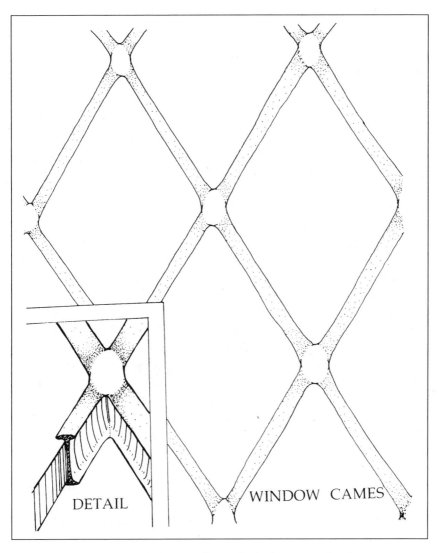

DETAIL

WINDOW CAMES

21. Section of window lead showing the method of securing the glass

scriptions, and in England the number was even lower. Cames of the type which bear inscriptions were made in a mill especially designed for the purpose, introduced sometime during the sixteenth century. Before their appearance cames were cast and, as a result, much heavier and thicker in section. Milling made it possible to produce thinner cames, with a high degree of control over their weight and dimensions. The process was a simple one, involving two wheels which rotated in such a way that their edges nearly met. The lead was fed into the space between the edges of the wheels, and a continuous strip of came was generated. The wheel edges were milled with transverse grooves which gripped the lead as it was fed into the machine. This in turn created reeding on the inside of the cames, visible on any example made using this process.

The makers engraved their names or initials and dates in reverse on the milled edges of the wheels, and as a result the inscriptions appeared on the cames. On pieces where inscriptions are repeated, they occur every five inches or so, which suggests a wheel diameter of one and one half inches. Although cames produced in such a mill had the advantage of being lighter and thinner than the earlier cast ones, there was an attendant danger, that of making them too thin and thus not able to support the weight of the glass which they held. The Glaziers Guild of London several times enacted fines for anyone making cames which did not meet standards of quality. Company officials were empowered to determine "whether any lead used had been drawn or extended beyond the proportioned length for the weight thereof," and the rules stated clearly that "any lead weighing one ounce and a length exceeding the length of one foot and three inches" would be substandard and liable for fine.[19] It follows that glaziers who wished their products to meet the standards would identify their work against failure, a seventeenth-century version of the name brands of today.

Unfortunately, these marked cames are less than ideal for dating purposes. Windows were shipped to the colonies in assembled form and installed as single units. Since we have no way of knowing how long they were stored in a warehouse before shipment, they do not necessarily indicate the date of construction of a house but could provide a *terminus post quem* for the event. On the other hand, a window might have been installed as a replacement for one that had been in a house built earlier, and thus the date given by the inscribed came would postdate the construction date. We are left then with a fascinating bit of information with no direct bearing on the question of the dates of occupation of site 98. But other artifacts provide us with more precise information, and the dated came at least agrees with this other evidence.

Which of the four structures, if any, is the dwelling house with a "good brick cellar" identified in the *Virginia Gazette* advertisement of 1767, the same building referred to as an ordinary by 1772? Only two possibilities present themselves, and one of them must be this building. Neither of the brick cellars was found by surface survey, plow zone testing, or remote sensing. Rather, their existence had been known for years before any archaeology was done at Flowerdew Hundred. Farmers had been striking cellar walls with their plows, and since the entire area has been under continuous cultivation at least during the present century, any other brick cellar on the site would have been discovered in a similar way. We must then determine which of the two once belonged to Joseph Wilkins. Fortunately, both the architectural remains and artifacts are quite clear on the matter. The little dwelling house at the northern end of the line of foundations probably was built after the date of the first advertisement. There are the few odd sherds of salt-glazed stoneware in the collection from its cellar, but this type of pottery was produced into the 1770s. The excavators of the cellar date it to the last quarter of the eighteenth century, and this seems entirely reasonable. The house which stood over the other brick cellar was certainly built before 1750, probably when the adjacent earthfast dwelling house was torn down, and dates from both structures provide mutual support for such a sequence. The ceramics from this cellar indicate an occupation into the 1770s, which would fit the second advertisement.

As the southernmost cellar was being excavated, there was a reluctance to identify it as the remains of an ordinary simply because it seemed too small to have served such a purpose. Furthermore, an uncritical reading of the 1767 advertisement made it appear to have had three rooms on the ground floor and three above. However, a more careful reading of the advertisement can well lead to a very different configuration. The three rooms could well have been arranged two above and one below, or the reverse, although one large first-floor room with two chambers above makes better sense, especially if the building had started its life as a dwelling house, as the advertisement makes clear. An ordinary in rural Prince George County probably did not do a thriving business, unlike the large taverns in the colonial capital of Williamsburg. Eighteenth-century concepts of privacy were quite different from those that came later, and it was not at all uncommon for people in inns to sleep three to a bed with

22. Lesser Dabney house, Louisa County, Virginia

two or three beds crammed into a room. The little ordinary at Wilkins's Ferry could have easily held as many as twenty guests a night. By modern standards this may not be the "good accommodation" and "entire satisfaction" promised by Sherwood Lightfoot in 1772, but for their time it was probably so. Further evidence strengthens the identification. The abundance of pipe stems and stoneware mug fragments in the immediate area, the wide entrance to the cellar, its specially prepared floor, and the cooper's tool found lying on it are more than sufficient evidence to support the identification of this cellar as the "good brick cellar" beneath Joseph Wilkins's dwelling house, later to become Sherwood Lightfoot's ordinary.

What the building might have looked like is indicated by a remarkably well preserved frame house of the later eighteenth century in Louisa County, some sixty miles northwest of Flowerdew Hundred (fig. 22). It is often difficult to project an underground archaeological feature into aboveground space, and this little dwelling house allows us to do precisely that. While it may differ in some details from the building that once served as the Wilkins's Ferry ordinary, it most certainly is a close approximation of what such eighteenth-century buildings looked like. This little house, first measured and described by Henry Glassie in his work on middle Virginia folk housing, is typical of a whole series of vernacular buildings in the area. Its dimensions are identical to those of Joseph Wilkins's house, and additional space was provided by sheds originally attached to both ends of the building, although one of them has since been removed. These sheds rested on brick piers and once taken away would leave no traces for the archaeologist to recover. The Louisa County house is a one-story building with an inside stairway leading to a loft above. Whether the Wilkins house was two full stories is not known, but even a divided loft could have accommodated numbers of guests.

So while we can be certain that the Wilkins's Ferry dwelling house/ ordinary has been located, the identification of the other structures at site 98 is less than clear. Was the little building with its large chimney the "new kitchen" referred to in the advertisement? If not, did it serve the dwelling house just to the north? The difference in the north-south alignments between the two sets of buildings would suggest that it did, but sometime after the brick was removed from the wall, someone cut a pit into the edge of the cellar and filled it with an assortment of early nineteenth-century pottery. The most likely people to have done so would have been those living in the northernmost house, but if so, did they build another kitchen elsewhere? Only further excavation will resolve these questions, and until every structure on site 98 has been accounted for archaeologically, we will be at a loss for final answers. At present, the situation is like trying to do a jigsaw puzzle with only some of the pieces available; until all are assembled, the full picture will not emerge.

What is evident thus far is that the entire transition from earthfast building to fully framed construction on full brick cellars is played out here on a single site, and the dates conform closely to those suggested by Carson and his fellow authors in their article on impermanent architecture in the southern colonies. The excavations that have been carried out have hardly scratched the surface of site 98, for it covers some four or five acres littered with artifacts which turn up with each year's plowing. We have seen that the elusive Powhatantown has never been identified through archaeological survey, and indeed a town in the modern sense does not appear ever to have existed at Flowerdew Hundred. On the other hand, the records say only that what actually stood at Powhatantown was "several warehouses and dwelling houses." Such a group of buildings could easily be accommodated by the five acres of site 98, which suggests that perhaps only one of the purchasers of land in the town developed anything substantial on the property. The association with a ferry indicated in the records, as well as the wine bottle seal bearing the name of one of these investors, John Hood, at least suggests that like Peirsey's rail, we may have already found Powhatantown without realizing it. Further support comes from names on the landscape, for Fort Powhatan stood just to the south across Flowerdew Hundred Creek, and Hood also had property in that area. The final answer to these questions must await further excavation, and site 98 is so large and complex, that it will be years before we know just what once stood on this part of Flowerdew Hundred.

By the time of the Revolution and into the 1780s, there were probably fewer dwelling houses on Flowerdew Hundred than at any time in the colonial past. Only four sites are known from this period. There were still people living at the Wilkins's Ferry site, represented by site 98. Colin Cocke had built his plantation house, which he named Belleview, at the southern end of the plantation, on the ridge overlooking the bottom-lands, and William Poythress had constructed a frame house on the same ridge some distance to the north. This house, represented by site 113, was the one that passed from the Poythresses to Mary Peachy and then to Miles Selden, and while it appears to have been built sometime in the 1780s, its significance to our story lies in its brief occupation in the early years of the nineteenth century. There is a fourth house in this set, but it is not located within the present boundary of Flowerdew Hundred Farm and is presently not available for archaeological study. This last site is the house of the Poythress family, located off the northern end of the property. Its location had been determined, but at present we have no knowledge of its construction date or period of occupation. It was destroyed by a fire in 1800. Such then was the setting at Flowerdew Hundred as the nineteenth century was ushered in, and the first truly American society made its appearance in the Chesapeake. The great conflict of the 1860s was still a half century in the future, but events equally significant to the people living at Flowerdew took place in the opening decades of the nineteenth century.

Chapter Five

THE VIEW THAT Miles Selden enjoyed from his veranda in 1810 was impressive. His house was perched on the edge of the ridge almost in the exact center of the thousand acres that had once been Flowerdew Hundred, and he could see the river beyond the ferry crossing and the lone dwelling house and kitchen there. To the north John V. Willcox's new house was visible, and to the south, Colin Cocke's stood on a slight rise. These four dwellings were the only ones standing on the property. The bottomlands were in full cultivation, and Selden probably never realized that the crops covered the remains of almost two centuries of human presence. Yeardley's and Piersey's settlement had long since vanished, as well as all of the other buildings that once dotted the fertile bottoms in the later seventeenth and eighteenth centuries. Life at Flowerdew Hundred was now confined to the ridge west of the river, and it was here that the final episodes in the Flowerdew story were played out. This ridge is actually the rim of a terrace, dropping some forty feet to the fields that border the river but remaining at the same elevation to the west. The soil here is different from that in the bottoms, stiff and full of clay. It was cultivated, however, and in recent years it has been cleared of secondary forest and planted once again. Selden's house was accompanied by a kitchen, barn, smokehouse and icehouse, and at this set of buildings we take our first look at what was taking place at Flowerdew Hundred in the opening decades of the nineteenth century.[1]

Before any archaeology was done, the existence of the house and its

No. 198.

Declaration for Assurance.

I THE underwritten William Sam.l Peachy residing at Flowerdehundred – in the county of Prince George do hereby declare for Assurance in the Mutual Assurance Society against Fire on Buildings of the State of Virginia, established the 26th December, 1795, agreeable to the several acts of the General Assembly of this state, to wit:

My three buildings on my Plantation called Flowerdehundred now occupied by my self situated between the Plantation of Collin Cocke – and that of Susanna Cocke in the county of Prince George their dimensions, situation, and contiguity to other buildings or wharves what the walls are built of, and what the buildings are covered with, are specified in the hereunto annexed description of the said buildings on the plat, signed by me and the appraisers, and each valued by them as appears by their certificate hereunder, to wit.

The Dwelling house marked A at 2300 Dollars, say Two thousand three hundred Dollars.
The Kitchen do B at 400 do four hundred
The Barn do C at 250 do Two hundred and fifty do

2950

say Two thousand Nine hundred and fifty Dollars in all.

I do hereby declare and affirm that I hold the above mentioned buildings with the land on which they stand, in fee-simple, and that they are not, nor shall be insured elsewhere, without giving notice thereof, agreeable to the policy that may issue in my name, upon the filing of this declaration, and provided the whole sum does not exceed four-fifths of the verified value, and that I will abide by, observe, and adhere to the Constitution, Rules and Regulations as are already established, or may hereafter be established by a majority of the insured, present in person, or by representatives, or by the majority of the property insured represented, either by the persons themselves, or their proxy duly authorised, or their deputy as established by law, at any general meeting, to be held by the said Assurance Society. Witness my hand and seal at this Twenty Sixth day of July 1801.

Teste Lewis Rivalain

W. S. Peachy (Seal)

WE the underwritten, being each of us house-owners, declare and affirm that we have examined the above mentioned property of Mr. Wm. Saml. Peachy and that we are of opinion that it would cost in cash Four thousand Dollars to build the same, and is now (after the deduction of One thousand Dollars for decay or bad repair) actually worth Two thousand Nine hundred and fifty Dollars in ready money, as above specified to the best of our knowledge and belief, and he the said subscriber has acknowledged before us his above signature.

Colin Cocke } Residing in the County of Prince George
Wm. Eyres

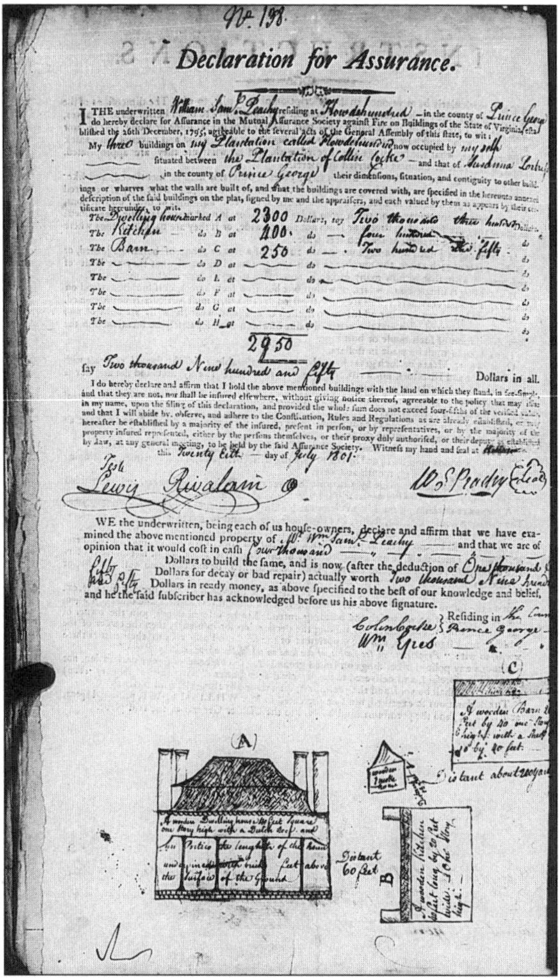

23. 1801 assurance policy for Miles Selden's house, site 113

outbuildings was known from three insurance policies, dated 1801, 1810, and 1816 (fig. 23). These documents are quite detailed; one even includes a sketch of the elevation of the house, and all three show plans of the various buildings and the distances between them. The house is described in two of the policies as "a wooden Dwelling house forty feet square, one story high with a Dutch roof and a portico the length of the home, underpinned with brick [*illegible*] feet above the surface of the ground" and "a dwelling house of wood, covered with wood, one story and hipped roof, forty by thirty two feet, with a British cellar." The difference between

the dimensions given in the two descriptions is accounted for by the portico, which is described as being ten feet wide, an error of only two feet. Two of the three policies give the dimensions as thirty-two by forty feet. The elevation drawing is quite detailed and agrees with the other descriptions and plans, showing the hip roof clearly and four chimneys, two on each end of the house. A British cellar means an English cellar, with brickwork rising above ground level far enough to accommodate windows.

The archaeologists' task in this instance was relatively easy, simply to locate and excavate a building that would fit the description given in the policies. It was known from early site survey that a substantial site (number 113) dating to the proper period was located on the property owned by William Peachy and later sold to Miles Selden in 1807 by his widow Mary. Excavations were begun there to get some idea of what the house might have looked like and to learn what artifacts were in use at Flowerdew Hundred in the early nineteenth century. These reasons for conducting the excavation were a bit different from those that have motivated other archaeological work on the farm, being more museological in nature, with no strongly developed set of research questions in mind at the outset. But as so often happens, what we found is not what we set out to discover, and in the course of two seasons of excavations at the site, other questions emerged and once again called for comparisons with other materials located far from the lower James River valley; far indeed, for before the last *i* is dotted and *t* crossed, we will have need to consider sites and buildings located some eight thousand miles away, at the southern tip of the African continent.

But we must begin at the Selden house, for it is there that the initial discoveries took place. The area in which the site survey identified a rich surface scatter of late eighteenth-century and early nineteenth-century artifacts had been in forest until the late 1960s. Logging activities had torn the ground up rather badly, and since no survey had been carried out before the area was cleared, we have no way of knowing what the site might have looked like before that time. But there are some clues that the cellar may well have been only partially filled and still visible in general outline even at that late date. Very early in the course of excavating the cellar, it became obvious that the Selden house had been found. When the corners of the cellar were located, they measured exactly thirty-two by forty feet apart, and artifacts supported the dates of the final occupation. There was no evidence of the veranda, which stood on the river side of the house, but this is not surprising, since it probably was supported by brick piers whose remains would have been obliterated by the logging. The top two feet of the cellar seemed to have been filled when the area was cleared, for oilcans, wire, and other modern debris were found only a scant foot above the floor in the center, although the earlier fill was deeper along the walls. This evidence suggests that the cellar hole would have been visible before the area was cleared. Standing among trees, the cellar would have filled very slowly; New England abounds with cellar holes from the nineteenth century scattered through the forests and clearly visible as shallow depressions filled with leaves. Logging activity not only obliterated much of the surface traces of the buildings on the site but also seems to have erased any evidence of the substantial kitchen. Extensive excavation in the area where the kitchen was known to have stood from the maps revealed only brick smears, the faint remains of the chimneys that once stood on either end of the building and are shown on the one policy in elevation. Only features that were cut deep enough below grade have survived on the site, and two such features were excavated, the cellar and an adjacent icehouse.

When fully excavated, the cellar showed an unusual floor plan (fig. 24). If the cellar partitions reflect the layout of the rooms that once were above, as is almost always the case, the floor plan of the house does not conform to that which was expected from the elevation drawing, which showed paired chimneys on either end, suggesting a house two rooms deep, with pairs of rooms flanking a central hall in the standard Georgian layout. The plan is one of four rooms, and the bases of the chimneys were located where the drawing showed, but the house apparently lacked a central hall. Entrance was into a large room with a smaller room to the left. The pair of rooms to the rear were also of unequal size, but reversed, with the smaller room to the right. Entrance to these rear rooms probably was through either a single door in the rear of the larger front room, or doors in both front rooms leading to the rear. An almost identical floor plan can be seen at Shirley Plantation, the house built by John Carter across the river in ca. 1738–39. At Shirley doors lead from each of the front rooms into the rear, a layout much appreciated by tour guides when taking groups of visitors through the house. But neither Shirley nor the Selden house was designed to benefit the touring public, and the reasons for the floor plan at the latter much be sought elsewhere.

The house was constructed well before Selden owned it, and the most probable person to have built it, and perhaps resided in it, is William Poythress, who acquired this parcel of land in 1782, although it is possible

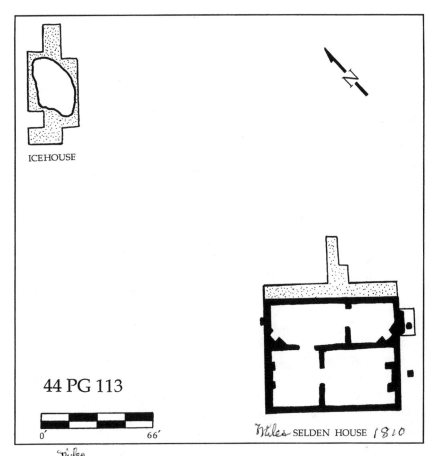

ICEHOUSE

44 PG 113

0′ 66′

Miles SELDEN HOUSE 1810

24. Plan of Selden house cellar, site 113

that it was constructed by James Warthan, from whom Poythress purchased the property. This uncertainty is due to the dates suggested by artifacts from the cellar floor. While most of them, ceramics and wine bottles in particular, suggest a date after the third quarter of the eighteenth century, there are a few pieces of Staffordshire salt-glazed stoneware that could have made their way into the cellar at an earlier date and a silver "mote spoon," used to strain tea, which is of a George I design. However, silver objects were highly valued and, as such, are not in the same category as everyday ceramics, making them of less value for dating purposes. Most of the ceramics were creamware and pearlware, types characteristic of the later eighteenth and earlier nineteenth centuries, so the house could well have been constructed after 1782. In any case, it was standing well before it came into Selden's hands, and the final date of occupation indicted by the artifacts appears to have been sometime in the late 1820s or early 1830s. Selden died in 1814, and the latest insurance policy (1816) states that the house is in the care of an overseer. But evidence from the icehouse nearby challenges this scenario, as well as raising new questions.

The icehouse almost wasn't excavated, for its existence was known from neither records nor artifacts on the surface. On a busy July morning at the site, the excavation of the cellar has progressed to detail work on the floor and mapping of the walls, and there were suddenly more people than places to work. In order to accommodate this labor surplus, a group was assigned to investigate the slightest of depressions to the rear of the house. This depression was so faint and shallow that it was better felt through the feet as one walked over it than seen. Two ambitious trenches, each ten feet long, were laid out at right angles with the depression at the center. Nothing much was expected to turn up, but it did occupy several people usefully at the time. At first little appeared, just the odd pipe stem and occasional small scrap of pottery. The hard clay of the ridge was particularly unyielding, and the effort was about to be abandoned when the excavators encountered a large groundhog burrow inside of which they could see a dense mass of brick and a large metal object projecting from one side. Indeed, the depression may well have been made by the burrow, which had collapsed in its uppermost section. This discovery led to an excavation that would occupy workers the entire season to follow, revealing the remains of an icehouse fifteen feet square and fifteen feet deep, filled with an amazing variety of objects dating to the first years of the nineteenth century (fig. 25).

Icehouses were common structures in the colonial period and are in use in remote parts of the country to this day. Ice was cut from a stream or pond during the winter and stored in a deep pit which usually had a drain at the bottom to carry off meltwater. The ice was layered with sawdust which acted as insulation. A pit the size of the one at the Selden house could easily have held enough ice to provide an ample supply during the summer months. The ice probably was cut from a pond nearby which had been created by damming a small watercourse in the vicinity. Icehouses can be either round or square and range in size from eight feet or so across to truly immense pits, such as that at Westover Plantation across the river which is over thirty feet deep. The Selden icehouse prob-

25. Selden icehouse during excavation, site 113

ably was lined with brick originally but had been robbed of its bricks sometime before it was filled with refuse. It had been dug into clean porous sand, which acted as an efficient drain for meltwater runoff. When excavators reached the bottom of the pit, and water was run into it in a heavy flow, the water disappeared into the sand as fast as it poured in.

The first task the excavators of the icehouse undertook was to cross-section what at first looked like a shallow circular depression, some eighteen feet in diameter, filled mostly with brickbats and dumped mortar, as if someone had been doing masonry and was disposing of the leftovers from each day's batch. The bottom of the pit appeared to be clean, undisturbed clay, and the groundhog burrow stopped short of cutting through it. Artifacts from this upper level of the icehouse tell us that there had been people doing something on the ridge in the earlier seventeenth century. They were not all that numerous, but in sufficient numbers to arouse one's curiosity as to how they might have gotten there. Of course, in historical archaeology it is not all that unusual to encounter fill that has been moved a considerable distance from its point of origin. A good example of this is provided by a block of South Main Street in Providence, Rhode Island, where eighteenth-century cellars were filled during the nineteenth century with earth hauled all the way from Attleboro, Massachusetts. But that was done in an urban context, and it is difficult to imagine the circumstances that would have led to fill being carried all the way up from the river bottoms where all of the known seventeenth-century sites are located. However, the artifacts found their way into the icehouse fill, and their dating is clear and unambiguous; pipe stems with bore diameters of $8/64$ inch, Chesapeake pipe fragments, a mask from the neck of a bellarmine jar, and a large piece of a delftware ointment jar dating to the first years of the eighteenth century.

Most intriguing were two pieces of a French burr millstone, suggesting that someone was milling something not too far away. French burr was a highly prized material for making millstones during the seventeenth century to process grains exclusively for human consumption. Millstones made from this material were composite affairs, for French burr did not occur in a form that would permit a millstone to be made from a single piece. The pieces were joined together and bound with iron straps, rather like iron wagon tires. The discovery of these two millstone pieces led quite naturally to the question of where the first windmill at Flowerdew Hundred was located. Conventional wisdom had always placed it somewhere near the Yeardley-Peirsey settlement, partly because of the name Windmill Point given to the tip of land jutting out into the river there. But anyone who has spent even a week at Flowerdew Hundred knows well that there are many days when the air along the river hardly stirs, while along the ridge to the west there is almost always enough breeze to drive the vanes of a mill. David Harrison has had a memorial post-mill constructed on the ridge only a short distance from the Selden site, and there are few days when the wind is not sufficient to drive the mill with all of its heavy wooden gears and massive millstones. This mill was sited for reasons that had nothing to do with the possible location of the mill built in 1621, but it may well be that it is nearer to the location of that first mill than we had thought, if the pieces of French burr millstone in the icehouse fill came from nearby.

After the top layer of brickbats, mortar, and artifacts was removed, the excavators thought that the feature had been fully defined. But just to be certain, a small test was made in the center of the depression, and only a foot below, sealed by a layer of clean water-laid sand, they broke into a solid mass of refuse: bricks, broken plates, bottles, and drinking glasses, masses of animal bone, tools, smoking pipes, eating utensils, and a host of other objects, looking for all the world as though someone had tipped a house on its side and allowed its contents to pour into the gaping hole in the ground. Down they dug, layer after layer, and finally, fifteen feet below the surface, the bottom was reached. But in removing this remarkable mass of material from the earth, spreading it out, and looking at it with a critical eye, one was forced to ask, Why was all of this stuff put into the ground at the same time; what event would lead to such mass disposal of apparently perfectly usable goods?

There is no question that the icehouse pit was filled quite rapidly; it was not a deep convenient pit that slowly accumulated refuse from the house nearby but rather contains a mass disposal of a variety of goods that might well have taken only a day to complete. Evidence for this rapid filling is provided by cross mends between numbers of ceramic pieces which tie the topmost level with that at the very bottom. There were visible strata in the fill, but these can occur when material from different locations on the site is dumped in a very brief time. That the material in the pit came from the house is shown by numerous cross mends between a variety of ceramic pieces in the fill and other fragments on the floor of the cellar. The icehouse fill contained large quantities of animal bone and oyster shell, suggesting that cleaning of the house lot accompanied the filling of the open pit. There is also evidence of some kind of remodeling, with both construction (the brickbats and mortar dumps in the top layer) and destruction (whole brick with mortar adhering) indicated. But the most telling evidence suggesting that this refuse was not the result of day-to-day breakage is the large number of restorable pieces of pottery, stacks of plates, rows of earthenware and porcelain bowls, mugs, and cups, as well as numerous bottles that could be pieced back together.

What are we to make of such a rich deposit of objects that almost certainly were broken in the process of disposal? Combined with evidence of cleaning up the house lot and some kind of construction activity, it would seem that someone was fixing things up, cleaning, building, and replacing one set of objects with another. The most immediate explanation would be to suggest that filling the pit, cleaning up the property, and perhaps doing a little remodeling related to a major event in the life of the residents of the house—a death, marriage, or transfer of property. The *terminus post quem* for the icehouse fill is between 1825 and 1830, and the property was purchased from the Selden estate by the Willcox family in 1827. It also seems very likely that someone was living in the house after it was acquired by the Willcox family, although they continued to reside in their house to the north. Possibly the overseer mentioned in the 1816 policy lived in the house into the 1830s, but since it vanishes from the records after 1816, we will never know. The house may have been standing during the Civil War since it is shown on a map from that period, with the name Selden still used as a designation, along with the names Cocke and Willcox identifying the other two houses on the ridge. However, it is possible that this map, though used in the Civil War, was drawn earlier.

The most important question that must be addressed in using major life events for explaining a dump like this is why such deposits have not been found anywhere on American sites which date either earlier or later than the third decade of the nineteenth century. People have always sold property, gotten married, and died, yet the earth has not yielded evidence of these events in the form of mass dumping. Furthermore, the icehouse at Flowerdew Hundred stands in company with at least a dozen others, excavated along the eastern seaboard from New Hampshire to Georgia. When the contents of these dumps are compared, it becomes instantly clear that all of them date to almost the same time, have very similar characteristics, and thus must relate to some larger change in American society.

By far the most spectacular mass dump was found at the Narbonne site in Salem, Massachusetts. The Narbonne house, which still stands on the site, was constructed in 1670 and was continuously occupied through the eighteenth, nineteenth, and early twentieth centuries. The house sits on a narrow, deep lot, and the archaeologists exposed almost the entire property, so that virtually all refuse disposed of on the site during its entire period of occupation was recovered. Before the 1770s refuse was simply broadcast across the lot, forming what is called sheet refuse. Various shallow pits on the lot were filled with refuse as well, but these were not dug expressly for that purpose. During the last quarter of the eighteenth century, this manner of disposal gave way to disposal in pits specially constructed for the purpose. Geoffrey Moran, the archaeologist who directed the Narbonne excavations, has divided the occupational history of the house into periods defined by the various families who resided there.[2] He presented the artifact yield for each of these periods in tabular form (table 2). Even a quick glance at this table confirms the point made earlier, that major life events have no relationship to the kind of dumping that we see at site after site dating to the third decade of the nineteenth century. The Narbonne site is the "smoking gun" in the argument set forth here, since all periods are accounted for through the complete excavation of the house lot, and only during the period of the Andrews family's residence do we see the vast quantities of ceramics and other artifacts being disposed of on a wholesale scale, along with architectural remains such as those seen at Flowerdew, all placed in wood-lined pits dug especially for the purpose. The Narbonne site was excavated in 1973–75, and at the time the spectacularly rich trash pits were not perceived as being representative of a much more widespread pattern of mass dumping. Yet Moran in his reporting on the site refrained from attributing the mass disposal to a significant event in the lives of the Andrews family but rather suggested that an emerging "style consciousness" and "concern for amenities" led to the disposal of so many perfectly useful objects. He connects this to the remodeling of the house at the same time, as well as the construction of a carriage house. In this explanation Moran came very close to saying that such mass disposal of goods had something very basic to do with people's attitudes and self-perceptions.

What are the specific characteristics that tie all of these dumps together? First, there are the ceramics. Two types make up the vast majority of the pottery in them. Creamware is, as the name suggests, a light cream color, almost white, and was the first truly mass-produced pottery to be made at the Staffordshire potteries. Its invention is attributed to Josiah

Table 2. The Narbonne house trash pit contents by period

	1700–57 Willard		1757–80 Hodges		1780–1820 Andrews		1820–70 Narbonne	
	No.	%	No.	%	No.	%	No.	%
Ceramics*								
Class 1	288	70.6	928	67.6	5,074	39.3	913	40.5
2	41	10.1	197	14.3	168	1.3	26	1.1
3	41	10.0	106	7.7	234	1.8	11	.4
4	18	4.4	21	1.5	178	1.3	17	.7
5			23	1.7	2,803	21.7	87	3.8
6			6	.4	2,874	22.3	375	16.6
7			1	.1			259	11.5
8	10	2.4	59	4.3	745	5.7	141	6.2
9	10	2.4	32	2.3	808	6.2	423	18.7
Total ceramics	408	24.8	1,373	32.0	12,884	33.6	2,252	34.7
Tobacco pipes	122	7.4	195	4.5	707	1.8	109	1.6
Glass	80	4.9	598	13.9	7,073	18.4	949	14.6
Building materials	195	11.8	395	9.2	3,068	8.0	865	13.3
Misc. iron	22	1.3	27	.6	1,820	4.7	936	14.4
Faunal	809	49.1	1,663	38.8	12,592	32.8	1,282	19.7
Other	10	.6	34	.7	154	.4	82	1.2
Total	1,646	99.9	4,285	99.7	38,298	99.7	6,475	99.5

Source: Geoffrey Moran, Edward Zimmer, and Anne E. Yentsch, *Archaeological Investigations of the Narbonne House, Salem Maritime National Historical Site* (Boston: National Park Service, 1982), p. 174.

*Ceramics classes: Class 1: redwares; 2: trailed, combed, and dotted wares; 3: delftwares; 4: fine white salt-glazed wares; 5: creamwares; 6: pearlwares; 7: hard white wares; 8: porcelain; 9: others.

Wedgwood, who introduced it in the mid-1760s. It remained popular through the early years of the nineteenth century, being slowly replaced in people's favor by the second type of pottery, known as pearlware. Pearlware is little more than creamware that has had traces of cobalt added to the glaze, giving it the faintest of bluish hues and making it appear whiter than creamware, the same effect that is obtained through using laundry bluing or soap powders with "blue magic whitener." Pearlware became immensely popular during the last years of the eighteenth century and remained so through the 1830s, when it in turn was slowly replaced by pottery much like we use today. Within each of these broad classes of pottery there are numbers of types defined on the basis of decoration.

The decorative types of pearlware have a direct bearing on the dating of mass dumps along the eastern seaboard. The commonest type of pearlware at the beginning of the nineteenth century is known as shell edge and occurs in both green and blue. As the name implies, the rims of various vessels are painted and at times molded in a way reminiscent of the edge of a shell. In addition to shell-edge wares, pearlware also appeared in forms that were painted, both in blue and polychrome, or decorated by a process known as transfer printing, in which a design is transferred with damp paper or gelatin from an engraved plate whose lines have been filled with pigment. Pearlware mugs, cups, and bowls were also often decorated with horizontal bands and are called annular ware. Shell-edge decoration; painting, particularly in polychrome using gentle, almost pastel, colors; and annular decoration are typical of pearlware used in the United States during the first decade of the nineteenth century. While transfer printing was invented in the mid-eighteenth century, its presence in quantity on American sites almost always suggests a date in the late 1820s or early 1830s.

The pottery from all of the mass dumps thus far excavated is characterized by a mix of pearlware and creamware, with pearlware represented by a slender majority, usually on the order of 60 percent or so. Within the pearlware category, shell-edge, annular, and painted types are predominant, with few transfer-printed pieces represented. This ceramic evidence suggests a date sometime in the first fifteen years of the nineteenth century for the dumps, but in most instances there are the odd and significantly small pieces of transfer-printed ware, later molded shell edge, or dated teacup handles that indicate a *terminus post quem* some ten years or so later than the date of the preponderant majority of the ceramics. The fact that these fragments usually do not represent restorable pieces suggests that they found their way into the deposits as the result of routine breakage, for under such circumstances one rarely recovers all of the pieces of a vessel. The dumps seem to be massive deposits of objects which are about ten years older than the date of their deposition. This indicates that people were clearing out their houses of older ceramics to be replaced by newer services, perhaps nearly the same as those thrown away. Other material from the dumps tells us that they were remodeling their houses at the same time and, in some cases, cleaning up the house lot as well, if the animal bone is interpreted correctly. These distinctive features are shared by all of the deposits of this type that have been excavated to date. These include, but are not limited to, the following sites.

Excavations on four domestic house lots in Portsmouth, New Hampshire, revealed privies, a well, and a trash pit that appear to have been filled rapidly in the early nineteenth century.[3] The bulk of the material from these features consisted of glassware, ceramics, and architectural materials suggesting remodeling. The excavators stated that they recovered "hundreds of reconstuctable tea service vessels" and "stacks of blue edge [shell edge] plates." A single filled cellar on the Hart-Shortridge property contained over four hundred reconstructible ceramic vessels. Many of the objects were dropped into the cellar whole, and some were even in stacks. A date no later than the 1830s is suggested for this material, but in view of the quantity of shell-edge pieces, the date of most of the material could well be somewhat earlier. Such was certainly so in the case of a pair of trash pits excavated on Main Street in Plymouth, Massachusetts. At the rear of a shoe store near the town center, two wood-lined trash pits five feet square and eight feet deep produced no fewer than 101 fully restorable vessels. When the site was first encountered, the archaeologists thought they were digging up damaged inventory from a store, but upon inspection, all of the dinner plates showed extensive wear from knives and forks. These ceramics included a set of ten matching Chinese export porcelain teacups and saucers, cream and pearlware plates, annular ware bowls and pitchers, and a variety of locally made red earthenware pieces. Once again, the majority of the pearlware is more typical of the first two decades of the nineteenth century, but the *terminus post quem* is provided by a transfer-printed saucer in the familiar willow pattern, produced in a Staffordshire factory that did not begin making pottery until 1822. A site known as "Black Lucy's Garden" in Andover, Massachusetts, also contained a dump of this type.[4] There a single small cellar produced 113 restorable vessels, which fit with those from other sites in terms of their type and dates. Pearlware vessels accounted for 51 of these pieces, and creamware, 35, the expected relative numbers for such dumps. The other vessels fell into a variety of other types, with only a few in each category. A series of privies in Alexandria, Virginia, yielded a spectacular collection of completely restorable creamware and pearlware pieces, and as in the other instances, they are almost entirely typical of those in use during the first twenty years of the nineteenth century. Other similar refuse deposits have been found at the Emerson-Bixby house in Barre, Massachusetts, in Middletown, Connecticut, and as far south as Camden, Georgia.

All of this evidence leads to a conclusion which needs an explanation. It is quite clear that sometime during the late 1820s or early 1830s, large numbers of people in the period of the early Republic threw out quantities of household goods and set about ordering their immediate world anew. It can be no coincidence that all of these dumps also suggest remodeling and most of them contain materials that indicate the people were cleaning and reorganizing the lot on which their houses stood. What was happening during the second or third decade to bring such events to pass? It is tempting to suggest some kind of religious motivation, since the period was that of the Second Great Awakening, a time when revivalism was strong, and a number of new sects made their appearance, including the Millerites and the Church of Latter-Day Saints. Yet just how these movements would bring about the reaction that the archaeology indicates is less than clear. But we can proceed along one line of reasoning and perhaps produce an explanation that, while speculative, accommodates the facts as they are now known.

To begin, site-specific explanations seem far too particularistic. Such explanations have ranged from a fear of sickness leading to mass disposal and replacement with "uninfected" objects, through their correlation with major life events, to the need to provide drainage at the bottom of the pits in which the refuse has been found. But such interpretations are not sufficient to explain why these features are all so similar and apparently were formed at almost the same time throughout the eastern United States. The religious argument has greater force in this respect, for these revivals affected large sections of the country at about the same time. But this explanation is a bit on the mystical side and lacks a clear connection between revivalism and mass disposal. Could it be that even a nation-wide perspective is not sufficient to produce an explanation that makes at least a minimal amount of sense? Comparative data from another part of the English colonial world would make this seem so, and we now must turn to the English frontier in nineteenth-century South Africa.

Looking at the same colonial culture as it manifests itself in a different environment can often produce a new way of thinking about material from each area in question. Such an exercise is little more than a process of controlled comparison. The culture of the 1830s in North America and South Africa share a common English parenthood, reflected in many aspects of the archaeology and material culture in both areas. The only significant variable then is that of geographical location, and specific aspects of the development of English culture on each of the frontiers can be considered in that context. Eastern North America was still a frontier of

English expansion in the early years of the nineteenth century. Although the colonies had achieved their independence twenty-odd years before, the culture of the colonies remained English well into the nineteenth century and even beyond in many of its aspects. The society that developed on the Eastern Cape frontier in South Africa after 1820 also was English, although over time it, too, slowly assumed a form that would set it apart from its mother culture back in England.

South Africa was settled by the Dutch in 1652 as a provisioning station for ships of the Dutch East Indies Company sailing from Holland to Batavia and back. By the end of the eighteenth century, settlement had spread from the Cape of Good Hope eastward to the Great Fish River, and when control of the colony passed to the English in 1806, the Fish River frontier was a volatile one. Here the Europeans first encountered great numbers of Nguni-speaking indigenous people, pastoralists who had been slowly moving south and west for centuries. Rather than develop a strong military presence along this frontier, which would have been prohibitively costly, the government devised a scheme whereby the area could be quickly settled by immigrants from England. Such a course of action also helped ease the serious unemployment that followed the Napoleonic Wars. Monetary incentives were offered and transportation to the Eastern Cape was provided, and in 1820 five thousand English people arrived in Algoa Bay and proceeded to settle in the region. These people, known today as the 1820 settlers, established an English colonial culture in the Eastern Cape that closely resembled its American counterpart in the early nineteenth century. The Cape remained an English colony, but compared to America, it was far more isolated by both neglect and distance. For these reasons, Eastern Cape culture and that of the United States slowly diverged from that of England, and this development can be traced in the material record of each area. One site in the Eastern Cape illustrates this process in microcosm, and both architectural research and archaeology have been carried out there.[5]

The site is known as The Hall, the name given on early maps to the house that stands there. The Hall is located in the small village of Salem, established as a Methodist community in 1820, some fifteen miles from Grahamstown, the major administrative and trading center of the district. The Hall grew over the years in a series of stages. Beginning as a small, one-room farmhouse built sometime in the 1820s, it was first remodeled into a hall-and-parlor house in 1834. At some later date this structure was enlarged by constructing a one-story central-hall I house on one end,

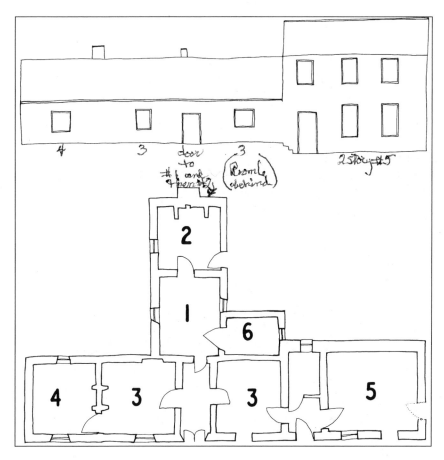

26. The Hall, Salem, Cape Province, South Africa

forming a T-shaped building. Later, probably around 1870, an even more elaborate addition was made to the house in the form of a two-story two-thirds I house on the end of the earlier one-story addition (fig. 26). This house has been thoroughly measured and its various construction stages examined to determine the various uses of space as these changed over time. Excavations have been carried out both in front of the house and, most significantly to our story, behind a section of stone wall to the rear. This wall probably once enclosed the entire house lot, and sections of it have been traced well beyond the short section which stands today. Behind the wall test excavations revealed a deep deposit of artifacts extending along its entire length.

When this deposit was excavated more extensively, the true nature of its richness became apparent. I codirected this excavation and the excavations of the Plymouth trash pits, had been present when the richest deposits were excavated at the Narbonne site, was codirector of the excavations at the Miles Selden icehouse, and had examined the material from Portsmouth and Alexandria in considerable detail. One need only to have encountered one or two of these mass dumps to appreciate them for what they are and to realize how different they are from other trash deposits. Once seen, they can be mistaken for nothing else. The deposit behind the wall was yet another such mass dump, with the usual fully restorable ceramics, bottles—many of them still with caps and seals—and a varied assortment of other household objects. As at the Selden icehouse, cross mends between the lowermost level and the top tell us that the deposit formed very quickly, and like similar American examples, the fill contained evidence of remodeling. Quantities of animal bone were also recovered, suggesting that the house lot was being cleaned up while the other materials were being disposed of.

Although a single dump in South Africa is not the same as a dozen or so in the United States, the similarities between it and its American counterparts are so close that they can be treated as a set, and a single explanation sought for all of them, based on what we know about the development of both colonial cultures. This explanation turns on the dates of the dumping in each case. While all of the American dumps appear to have been formed sometime around 1830, give or take a few years, the *terminus post quem* for the dump at The Hall is ca. 1870, based on marks found on plates and teacups. Yet the majority of the material from the fill is some ten to fifteen years earlier, as it is in the American examples, suggesting that the residents at The Hall disposed of older materials from the house as well. Test excavations in the front yard revealed an older land surface that had been raised and essentially leveled at about the same date. This landscaping and the dumping behind the wall almost certainly occurred when the large two-story addition was added to the end of the house. So we see that in both America and at least one site in South Africa, people were remodeling their houses, cleaning up the house lots, and replacing a significant portion of their personal possessions with new ones. While the dump at The Hall is the only one thus far excavated in the Eastern Cape, collections made by farmers elsewhere in the area and their description of dense concentrations of material on their properties strongly suggest that other households were undergoing the same kind of changes at about the same time.

How then might we account for these very similar events taking place, at such a great distance apart, and separated in time by forty-odd years? The answer probably relates to certain fundamental changes in the form of the colonial society in each case. The reordering of the people's physical world took place some fifty years after the initial separation between each frontier society and the parent culture. The 1830s are fifty years after the end of the American Revolution, and in South Africa the 1870s are fifty years after the arrival of the 1820 settlers. While the latter politically was still English, other factors contributed to separation. South Africa was a very different world from England, in terms of both environment and interaction with the indigenous society. Communication between England and the Cape was never strong, and after the first five thousand people settled on the Eastern Cape frontier, only a trickle of further immigration from England took place during this critical half century. Although America became politically independent in 1783, interaction between the former colonies and England remained at a relatively high level; but the Cape Colony, while still politically a part of England, was isolated to a much greater degree.

The end result was therefore quite similar in both places and led to the emergence of a distinctive local culture, still rooted in the English tradition but distinctively American or African as well. The years 1830 and 1870 were the time when the first generation of native-born Americans or English South Africans had reached maturity, and thus the archaeology seems to signal that critical point where the culture is no longer simply an extension of England, but rather American or South African. This explanation may seem a bit farfetched and admittedly is very speculative, but it does have the advantage of accommodating all of the facts at hand, and other bits of evidence provide a bit more credibility. In the Cape Colony a distinctive kind of architecture made its appearance at about the same time, characterized by a strict separation of space within houses between the residents and domestic servants.[6] This is seen at The Hall after 1870 in the relegation of all cooking activities to the older portion of the house, for the hearths in the two-story addition are small ones, used only for heating. In America the 1830s mark the spread of the Greek Revival style of architecture, a distinctively American expression. Recalling that religious revivalism was strong in America at the same time, although not being able to relate it directly to the reworking of the cultural landscape, we might suggest that it, too, was the result of this change in self-perception. It may not be a coincidence that a strong religious revival occurred in the Eastern Cape in the late 1860s and that Salem was particularly involved in it.

Comparison with South African vernacular building also allows us to look at the floor plan of the Selden house in a different way. Although from the exterior it looked like a typical Georgian house, two rooms front and back, separated by a hall, its interior arrangement did not follow that plan. The Eastern Cape region abounds with one-room-deep houses that appear from the outside to be central-hall I houses. Yet upon entering these houses, one usually passes into a single large room, with a smaller one off to one side or the other. As such, these houses exhibit a facade more in keeping with the concepts of privacy and individualism as shown by their American counterparts, while their interiors are far closer to the older hall-and-parlor house type. In America hall-and-parlor houses are typical of an earlier, preindustrial agrarian life-style, and in their layout they reflect the values of the public over the private and the group over the individual. However, by the time the 1820 settlers arrived in the Cape, central-hall I houses were common in both America and England. But when the settlers left England in the early nineteenth century, they also turned their backs on a world shaped by the industrial revolution and, upon arriving in their new land, turned with mixed success back to farming. Promotional materials used to recruit settlers in England spoke glowingly of the verdant lush countryside of the Eastern Cape, while in truth it was a harsh landscape, subject to long droughts and covered with plants they had never seen before, thorny bush and fat, succulent aloes and euphorbias. It took ten times the land to raise a given number of sheep there than it had back home in England. In spite of this new and hostile environment, the settlers managed to develop an agrarian society, with sheep raising the major economic activity. They built houses that looked modern and balanced from without, probably much like those they had known at home, but they changed back to the older hall-and-parlor layout on the inside, reworking a familiar house form to suit better the return to a rural way of life.

Other categories of material culture suggest the same shift to forms more typical of America in the eighteenth century. By ca. 1760 the American taste in ceramics had changed from a preference for colorful dishes to those of blue and white, and by the century's end slate had given way to marble as the preferred gravestone material, and epitaphs had become largely statements in the third person, stressing largely secular themes. At the Narbonne house and elsewhere, this was also the time when trash was carefully deposited in lined pits. The Eastern Cape settlers, on the other hand, preferred colorful ceramics, still available from the Staffordshire potteries but not popular in England and America. Josiah Wedgwood himself said that such colorful pottery was produced mainly for "the hot countries," and in the Cape transfer-printed pottery in a wide range of colors was used, at the time when American preferences were for mostly plain white wares, with some molded decoration. Eastern Cape gravestones continued to be made from slate well past the middle of the nineteenth century and retained first-person-singular epitaphs and strong religious themes. Refuse was still merely scattered about, producing sheet refuse more typical of eighteenth-century America and England. Clearly, the nature and direction of culture change took very different courses in the two colonies, and change in house layout is but one of a number of related transformations that occurred. The likely construction date of the Selden house makes it not entirely unreasonable to suggest that its builders were combining elements of Georgian design with older concepts of spatial layout, creating a house that looked to be in modern taste but would accommodate a life-style more comfortable and familiar to its occupants. But when John V. Willcox built his new house to the north, its room arrangement fully reflected the new order of balance, privacy, and individualism that had so completely transformed American society of the late eighteenth and nineteenth centuries.

The exact date of construction of the first dwelling house by Willcox is uncertain, although family tradition has it that he built a house at Flowerdew shortly after marrying Susannah Peachy Poythress in 1804 and coming into control of the property there. The first documentary record of a Willcox house at Flowerdew is to be found in the Prince George County Land Tax Books which record as a "sum added to Land on Account of Buildings" in 1820 an assessment for a $1,000 building on his 735-acre tract. But just what building is referred to here is less than certain. We have a solid cartographic anchor in the form of a coastal survey map made in 1857, showing a complex of buildings on the Willcox property (fig. 27). Included is a large dwelling house which has been traditionally identified as that built just after 1804. To the south of this building is a schoolhouse, which still stands today, and to the north, a kitchen beyond which is a line of shops, slave cabins, and, at the far northern end,

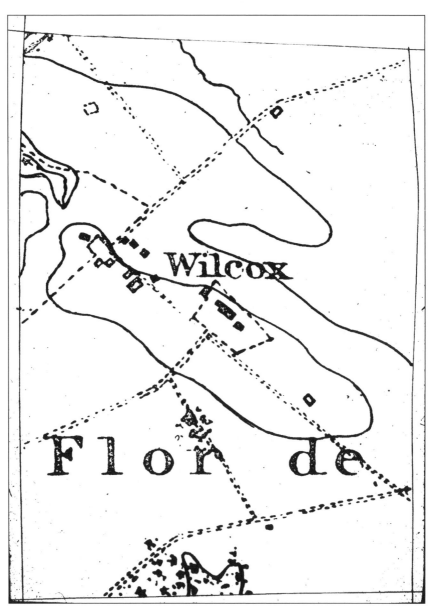

27. 1857 coastal survey map of the Willcox plantation. Slave cabins are the solid squares, *upper left*. Overseer's house, *farthest left*.

a structure identified on another map as the overseer's house. But whether the main dwelling house is the one that was constructed first by Willcox is less than certain; the little schoolhouse might well have been the first house built. When it was remodeled in 1982, exposed fabric was consistent with a date of 1804. Since the large house was razed in 1955 comparisons are impossible, but one has an intuitive sense that it may have been built considerably later, perhaps leading the tax assessor to amend his records in 1820, and that the schoolhouse may well have served as John and Susannah's first home. What is important is that by the 1830s a fully developed plantation community had been established on the ridge, and it was in the remains of two of these buildings that archaeological investigations were carried out. Both the kitchen and one of the slave cabins were fully excavated, and from them we get a glimpse of what life was like at antebellum Flowerdew and of some of the events that took place there.

The dwelling house was a standard central-hall I house, measuring forty-six feet by eighteen when first constructed. Porches and two rooms projecting to the east that were added during the later nineteenth century enlarged it to a total length of ninety-five feet (fig. 28). While no systematic excavations of consequence have been carried out at the site of the house, the Flowerdew Hundred staff prepared a detailed report on its architecture and construction sequence which provides a clear understanding of the building. There are ample photographic records of its interior and exterior, as well as oral history accounts from those who lived there. Such is not the case with the other buildings that once stood on the site. The kitchen is represented by only one photograph, and there is no pictorial record of any of the slave cabins or of the overseer's house. Archaeology thus provides us with our best knowledge of two of these buildings, one of four slave cabins and the kitchen.

Extensive testing failed to produce evidence of three of the four slave houses. The fourth was located and excavated completely, and since it stood until 1930, its remains had been less subject to destruction. If the other three were of similar construction, the inability to locate them comes as no surprise, because they were built in such a way that would leave minimal evidence of their presence below ground surface, and if they were razed at a much earlier date, and the materials removed for other construction, they would have left little or no trace of their presence. Furthermore, roadways in the area have been extensively realigned

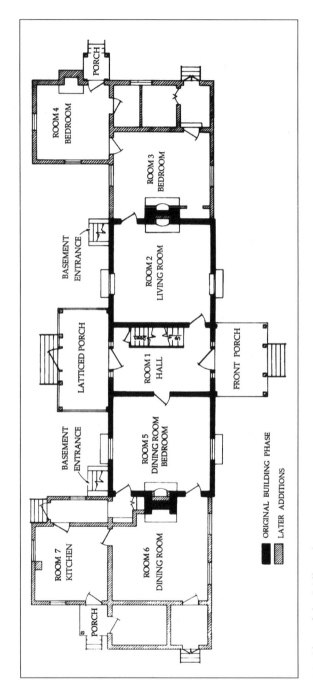

28. Plan of the Willcox house

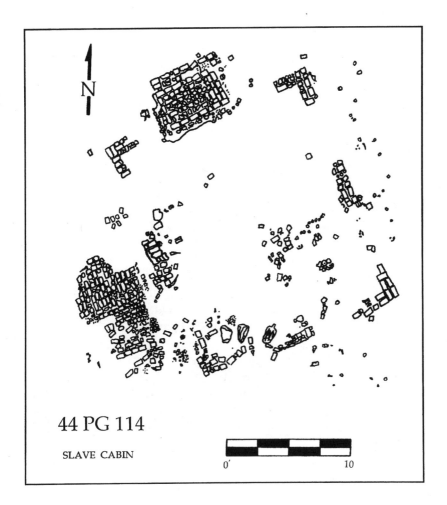

29. Slave cabin plan, site 114

during the twentieth century, and the coastal survey map suggests that the location of at least three cabins is under the roadbed today.

Excavations at the remaining cabin site provided evidence of a rectangular building measuring sixteen by twenty feet, raised above the ground on seven brick piers, with a chimney at the northern end. A rich deposit of artifacts and animal bone was located to the rear of the cabin on the eastern side. The front faced west, with entry through the long side, evidenced by a cobble-paved area (fig. 29). Comparisons with a number of standing slave cabins in Prince George County allow us to suggest what kind of building stood on the piers. It almost certainly was a one-room structure, with a loft overhead. The floor would have been of wood, at the level of the tops of the piers, producing a crawl space beneath the building. This space served two functions. It provided ample air circulation, which would reduce dampness and provide healthier living conditions. But the raised floor also tells us something about social control as employed by the planter. With a raised floor, it was not possible to dig "hidey holes," which are typical of eighteenth-century slave cabins excavated both at Kingsmill Plantation farther down the James and at Thomas Jefferson's Monticello. William Kelso, who directed both excavations, has interpreted these holes as places where slaves could conceal valuable possessions, perhaps including things not come by in legitimate fashion.

As a communal living space, the little cabin must have been quite crowded. The number of slaves owned by Willcox reached a maximum of sixty in the years before the Civil War, and if only four cabins stood on the plantation, this would mean that each would have housed an average of fifteen people. It is possible, of course, that additional cabins were constructed after the map was drawn in 1857, but such a figure is in line with what we know about crowded conditions in slave cabins on other plantations. The archaeology also provides other glimpses of what life may have been like for Willcox's slaves. Great care was exercised during excavation to separate materials from the antebellum period from those deposited later, and these shed some light on slave foodways at the time. While there was a variety of types of pottery, the most prevalent was annular-decorated pearlware bowls. Whether these were hand-me-downs from the Willcox family or purchased by them specifically for the slaves is not clear, although the larger numbers in the cabin as compared with the kitchen suggests the latter. Such a pattern was seen at Cannon's Point Plantation, Georgia, where the commonest ceramics from the slave quarter were annular bowls.[7] Annular ware was the second least expensive pottery available from England at the time, and it is encountered elsewhere in circumstances that suggest it was acquired by the dominant members of society for the subservient. It was the commonest type of pottery recovered from Indian neophyte quarters in La Purisima Mission, California (1812–36), and forms the major component of the ceramic assemblage from a South African frontier fort occupied by a "Cape Colored" regiment during the same period. Bowls as the commonest shape suggest a foodways pattern centered on stews and pottages, reminiscent of the earlier use of Colono ware bowls. What went into these dishes is evidenced by the animal bone and shellfish remains excavated from the deposit behind the cabin.

In the variety and quantity of wild animal bones in the deposit, the refuse at the Willcox slave cabin is consistent with a pattern noted by other archaeologists studying slave food remains on a number of sites in the South (table 3). All plantations maintained some kind of slave ration supply, and the surviving records tell what the planters were providing. What is missing from these sources is information on how much supplemental food the slaves procured through hunting and fishing. This is not to suggest that the slave diet was necessarily better than the records alone would suggest, but the archaeology provides a dimension to our view of slave subsistence that the records alone could never provide. While slaves at different plantations at different times obtained different wild animals, the essential fact is that regardless of the specific animals hunted, wild food supplies formed a significant portion of the slave diet everywhere. This in turn makes us consider the way in which time was managed by both slaves and planter so as to enable slaves to partake in hunting and fishing. The records indicate that most slaves maintained modest gardens, and that work was scheduled in such a way as to provide time to cultivate these small plots. From the planters' point of view this was advantageous in that it made the slaves less reliant on plantation rations. But the records are far less clear on the subject of hunting and fishing.

The archaeological evidence suggests that such time was available and well used. This is particularly so in the case of the large numbers of catfish bones at the Willcox cabin. These fish can only be taken with hook and line and thus require a significant investment of time. There is no way to tell from the archaeological record how often such wild food items made their way to the table, but their sheer quantities tell us that hunting and fishing probably formed a significant portion of the slaves' daily activities and made a contribution to the diet. In the category of domesticated animals, the commonest in the Willcox deposit is pig, not at all surprising in view of the popularity of pork in the present-day tidewater Virginia. The frequency of pork consumption probably was higher than the archaeology would indicate, since salt pork, without bones, was a common plantation ration item. The notable absence of some species of animals is of interest. Neither shad nor herring is present in the faunal sample from the slave cabin deposits. These fish occur in great numbers each spring as they migrate up the river to spawn, and they are easily taken. Larry

Table 3. Identified faunal materials from Willcox slave cabin, site 114

Species	No.	%	MNI*
Pig (*Sus scrofa*)	120	10.96	4
Cattle (*Bos taurus*)	13	1.19	1
Horse? (cf. *Equus* sp.)	1	.09	1
Sheep (*Ovis aries*)	4	.37	1
Sheep/goat	3	.27	1
Sheep/goat/deer	8	.73	1
Deer? (cf. *Odocoileus* sp.)	1	.09	1
Medium-sized mammal, unidentified	147	13.42	
Opossum (*Didelphis virginianus*)	12	1.10	2
Rabbit (*Sylvilagus floridianus*)	9	.82	2
Rat (*Rattus* sp.)	1	.09	1
Squirrel (*Sciurus* sp.)	1	.09	1
Raccoon (*Procyon lotor*)	1	.09	1
Mammal, unidentified	499	45.57	
Chicken (*Gallus gallus*)	29	2.65	4
Crow (*Corvus* sp.)	2	.18	1
Mallard/black duck (*Anas* sp.)	1	.09	1
Bird, unidentified	63	5.75	
Catfish (*Ictalurus* sp.)	81	7.40	21
Sturgeon (*Acipenser* sp.)	42	3.84	1+
Striped bass? (*Morone* sp.)	2	.18	1
Gar (*Lepisosteus* sp.)	1	.09	1
Fish, unidentified	45	4.11	
Snapping turtle (*Chelydra serpentina*)	4	.37	1
Turtle, unidentified	5	.46	
Total	1,095	100.00	
Shellfish			
Oyster (*Crassostrea virginianus*)	112		
Freshwater mussel (*Elliptio* sp.)	88		
Marine clam (species)	12		
Shellfish, unidentified	6		

Source: Lawrence William McKee, "Plantation Food Supply in Nineteenth-Century Tidewater Virginia" (Ph.D., Univ. of California, Berkeley, 1988).
*MNI: minimum number of individuals.

McKee, who excavated the cabin, and on whose dissertation this discussion is based, has suggested that the bones of these fish, being quite small and fragile, may not have survived in the ground.[8] There is independent documentary evidence of their being used as a food source at the Willcox plantation, so their absence from the deposit is puzzling; it seems that if they were consumed in any significant quantity, at least a few bones would have survived. Equally problematic is the complete absence of remains of the Canada goose. Today tens of thousands of geese spend the winter in the Flowerdew bottoms, feeding in the stubble fields. If such was the case in the nineteenth century, it seems that at least some small trace of their presence would have come from the slave cabin deposit. McKee suggests that hunting geese might have been forbidden by the planter, yet this seems unlikely if they were as common then as they are today. One other food source should be noted. Two species of shellfish were found in the deposit. One, the freshwater mussel, is locally available, although it has a very low food value. Oysters, on the other hand, which are also present in the deposit, are a more nutritional food source. Unlike mussels, they had to have been purchased by the planter for issue to the slaves, since the nearest source of them is some thirty-five miles farther down the James River, where the water has sufficient salinity to assure their survival.

The faunal materials from the Willcox slave cabin provide us with new insights on the subject of slave diet and subsistence. Taken with similar studies done elsewhere in the Chesapeake, they make a very important contribution to our emerging understanding of the subject. Furthermore, it may well be that careful analysis of the pattern of slave foodways as revealed by archaeology can help us resolve what is presently a vexing problem in Chesapeake archaeology. It is one thing to excavate a slave cabin identified as such from independent sources, such as the 1857 map of Flowerdew Hundred. But how does one define the ethnicity of the former residents of a site which can be seen to have been occupied by people of very modest means? Nothing less than the identification of unrecorded African Americans is involved here. At present, there are very few criteria that can be used in making such an assessment. But since foodways are often the most distinctive of ethnic markers, and conservative as well, further comparisons of faunal assemblages from a number of sites whose occupants, both European American and African American, can be identified might allow us to make this vital distinction on the basis of differences in the foods consumed and the manner in which they were

30. Detached kitchen, Willcox plantation, site 114.
(Courtesy of the Library of Congress)

prepared. But this has yet to be done, and in the meantime we can only continue to add to the knowledge of different foodway patterns at different levels of affluence and different ethnic affiliation, in the hope of resolving the problem.

Excavations at the kitchen, located between the main dwelling house and the slave quarter, produced no major surprises, although they did flesh out and provide texture to our knowledge of the building. The one photograph of it that exists is very good and shows a long structure, one room deep, two stories high, with chimneys at either end (fig. 30). When the building was fully excavated its exact dimensions could be determined, and certain aspects of its construction clarified. The fireplaces were of different size. The western one was much the larger of the two and apparently served as an all-purpose hearth for a range of cooking activities. The eastern hearth, smaller and of a different shape, may have contained an oven or possibly served as a fireplace to heat water for laundry, an activity commonly associated with kitchens in the region. The size of the building, the location of the chimneys, and the pattern of window placement seen in the photograph indicate that the kitchen was arranged in a two-room over two-room plan. The downstairs rooms almost certainly functioned as cooking areas, and probably for laundry as well, possibly with one room devoted to each activity. The upstairs rooms may well have been used as housing for slaves. If so, this space would have lowered the number of people in each of the slave cabins.

Just when the kitchen was built is less than certain. Sometime after the building was erected, a cross wall was added to the western end in front of the fireplace. This wall is not bonded to the exterior foundation, and its function is unknown. It was constructed in a shallow builder's trench, which contained a whole bottle typical of the 1830s, apparently purposely placed within. On the basis of this evidence, we can suggest that the building was constructed not later than the 1830s, and possibly much earlier.

The kitchen marks the place where the world of the Willcoxes and that of their slaves abutted one another. It is, in a sense, the mediation between the two worlds where the residents of the slave quarter, and possibly of the kitchen itself, worked to feed the planter and his family and to provide other amenities as well. This contrast is dramatically highlighted when one considers the pattern of refuse dumping seen around the kitchen. To the north and east sides is a vast and deep deposit of midden, while in the front and on the west side, both visible to the planter, there is no evidence of refuse dumping. The distinction here is between the ordered and manicured world of the planter, which enhanced his social standing, and the more cluttered world of the working plantation, which made his wealth and social standing possible in the first place.

Other artifacts from the kitchen excavations tell us that in later years it no longer served the purpose for which it was originally erected. Indeed, it probably did not serve as a kitchen for very long, if at all, after the Civil War. This pattern is seen throughout the region and tells us that once a supply of chattel labor was no longer available, separate kitchens went out of style, and cooking was incorporated into the main house. This pattern can be seen in reverse in South Africa at approximately the same time, when a supply of inexpensive labor came into being as more and more blacks moved into urban areas. There we see an emerging separation of living space from that involved in food processing and preparation. Likewise, in the early years of Australian colonization, a similar separation in space arose from the presence of large numbers of convict laborers. In all three cases, it is a matter of one class of people keeping a distance from the other, even though their services are critically needed for comfort, if not survival. Architectural separation thus resolves the problem of not wanting to associate with those people whose physical presence is essential.

In its final years the kitchen seems to have been the site of automobile maintenance. A wide variety of engine parts was recovered, and the latest dated artifact from any of the Flowerdew Hundred excavations, a Virginia license plate of the 1930s. This artifact speaks eloquently to the entire sweep of history at Flowerdew, for the archaeology has provided us with bits of material evidence extending from medieval styles of armor, chain mail, and crossbow parts in an unbroken sequence to the age of the automobile. Looking at all of the material that has come from the earth at Flowerdew, it is impossible not to react in an emotional way to the great changes it reveals, with each step along the way represented by some object or another.

When the Army of the Potomac crossed the James River at Flowerdew, a party of troops was quartered at the Willcox house for several months. Several stories have made their way into local tradition about this event. One story has it that the officer in charge cut the newel post of the stair in the main house with his saber. This may have happened, but the story bears a remarkable similarity to that of Carter's Grove, at Martin's Hundred, where the British officer Colonel Banastre Tarleton supposedly cut the newel post with *his* sword during the Revolution. Another tale has it that when the Union troops threatened to burn the Willcox house, the loyal slaves barricaded themselves inside and refused to come out, thus saving the house from incineration. These stories are probably just that— latter-day folk tales—but the kitchen excavations did produce evidence of the presence of northern troops. Minié balls, a bayonet, a cannonball, and two Union officers' belt buckles were recovered (fig. 31), as well as the chamber of an officer's pistol, with one possibly live round still in it. The cylinder was immersed in water to ensure that the last casualty of the Civil War was not some unfortunate excavator.

On June 13, 1864, after eleven days of some of the bloodiest fighting of the war, General Ulysses S. Grant withdrew from the strongly entrenched Confederate lines at Cold Harbor, just outside Richmond. His aim was to slip around Robert E. Lee's Army of Northern Virginia in a wide flanking maneuver, get to his rear, and cut off the lines of supply to the southern

31. Civil War artifacts from the Willcox kitchen, site 114:
Union army officer's buckle, minié ball, and cannonball

army. But one major obstacle lay in the way, the James River, a mile or more across in some places, one last river to cross between Grant and Petersburg, where the armies would meet again in their final tragic engagement. Engineers scouted the river for possible crossing places and settled on a spot where the river was relatively narrow, only half a mile across. On one side was Weyanoke Plantation, once owned by George Yeardley, and on the opposite shore, Flowerdew Hundred. In just one day a bridge of a hundred pontoons was thrown across the river, anchored on schooners in midstream. On June 14 the army made its crossing and moved across the Flowerdew bottoms, along the old road from the ferry, up the slight rise to the ridge, passing what may have been the still-standing ruin of Miles Selden's house. Local stories make reference to the troops using the ruin for target practice, but these are unsubstantiated.

While it was always generally agreed that the crossing took place somewhere on the Flowerdew property, the precise location was not known, although several locations were suggested, all at the lower part of the property, near site 98. It was not until the summer of 1988 that the exact site of the bridge was located with pinpoint accuracy. This accomplishment was made possible by a simple but ingenious device that is as close to a time machine as we will ever see.

Eugene Prince, photographer for the Lowie Museum of Anthropology at the University of California, Berkeley, had been experimenting with ways to incorporate historical photographs into various aspects of site location. By the mid-1980s he had come up with a technique that showed great promise using a camera equipped with a zoom lens, a tripod, a 35-mm transparency of the historical photograph in question, and a lot of inspired thinking.[9] The method is very simple. A description of the way it was used at the Willcox house will serve as a good explanation. Of the many excellent photographs of the house, one is particularly good, showing the house almost front on. For the technique to work, it was necessary to identify at least two points in the photograph on the modern landscape. More than two points would make the precision even greater. Using a single-lens reflex camera of the type with removable prism mounting, a transparency of the Willcox house was set horizontally on the surface inside the prism mounting, and the cover replaced. When one looked through the viewfinder, the photograph was visible and sharply focused. While a camera was employed in this technique, it was used only as an optical system; no photographs were taken. The next step was to look at the modern site and carefully superimpose the historical photograph on it, aligning the two reference points.

At the Willcox site these reference points were the corners of the foundation, still visible, and the corresponding corners of the house. A third point was provided by the center of the front entrance, still evident in the modern foundation. By moving around in front of the Willcox foundation and adjusting the focal length of the camera with the zoom lens, a point was found where the reference points in both the photograph and the modern site matched. The zoom lens was necessary because the focal length of the camera which took the old picture was not known. When one looked through the viewfinder, there was the Willcox house, exactly where it had stood in years past, a phantom house sitting on the old foundation. The effect was spectacular, for the photograph was actually tinted by the real world; the sky was blue, the grass was green. Once the camera had been properly sited, it was possible to direct a team of workers into the photograph and make measurements of anything visible, even though it is not there today. They actually entered the world of the past and could be directed by the viewer to relevant points from which to take measurements. The technique automatically corrects for perspective, since if a tape measure is run at an angle, it is at the same angle as the building being measured. The technique was put to very good practical use in the reconstruction of the Willcox kitchen, giving clapboard widths, window and pane sizes, and other dimensional details.

By the time the technique had been perfected, it was put to use in the search for Grant's Crossing, as that point on the river has come to be called. Gene Prince and Taft Kiser, then staff archaeologist at Flowerdew Hundred, scouted the river shore for likely locations, based on a Matthew Brady photograph of the pontoon bridge which prominently featured a large cypress tree on the near shore. There were several cypress trees that were good candidates, and one was particularly promising. When the camera with a transparency of the bridge inside was set up, and the proper distance and angle established, no fewer than nine points were found to correspond, almost all of them branches of the tree itself. It had changed over the years, to be sure, but certain key features remained, enough to establish beyond a doubt that this was the same tree as seen in the Brady photograph. The sight through the viewfinder was nothing less than breathtaking (fig. 32). There, stretching across the river, was the phantom bridge. Anchored in midstream were the wooden sloops of the Union navy, exactly where they had been on June 14, 1864. When the spacing between the pontoons was measured, it came out to be the fifteen feet recommended in the ordnance manual on the construction of pontoon bridges. Transit measurements of the point where the phantom bridge reached the other shore were within a few yards of the actual length of the bridge, known from historical sources. And when a test excavation was placed where the bridge reached the Flowerdew side, a roughly shaped square granite marker with a hole drilled in the center of its top surface was discovered. While we cannot say with absolute certainty that the marker was set by one of Grant's engineers, it is certainly not a modern survey marker, and the location seems beyond coincidence.

So it was that a puzzle was finally solved, using the simplest of equipment, but combined in a thoroughly ingenious fashion. The technique has been used to check the accuracy of a building reconstructed at Williamsburg, to establish certain details of building locations at Monticello, and to locate with greater precision the wagon park of the British army at the battlefield of Isandlwana, Natal, South Africa, where the Zulu armies inflicted the greatest defeat ever experienced by the British army. Its reach in time exceeds the introduction of photography in the mid-nineteenth century, for old photographs often show even older buildings that have vanished over the years. Such was the case at Fort Ross, a Russian settle-

32. Grant's crossing, Flowerdew Hundred, pontoon bridge of 1864
(*top;* from the National Archives) and modern site (*bottom*)

ment in northern California. There a shipyard was established in the early decades of the nineteenth century and remained in use long enough to have been photographed in 1862. These photographs were used to pinpoint the location of the shipyard buildings and launching way, all having long since disappeared. This use of camera along with historical photographs, dubbed "Prince's Principle" by Ivor Noël Hume, is a valuable addition to the tool kit of the historical archaeologist. As such, it is but one more way in which we have gained access to Flowerdew Hundred's past and created an absorbing story of one of the many strands that make up the history of both Virginia and America.

Archaeology at Flowerdew Hundred has indeed been a rewarding effort. The work of hundreds of people over nearly a quarter of a century has detailed the development and change on one Virginia plantation, an unbroken sequence from Yeardley's first tiny settlement on the banks of the James through to the establishment of Willcox's rather ambitious plantation in the mid-nineteenth century. The cast of characters is long and includes people whose names are known—George Yeardley, Abraham Piersey, Elizabeth Stephens, the Barkers and the Limbreys, the Taylor sisters, John Hood, the Poythress clan, the Wilkinses and Willcoxes, and finally General Ulysses S. Grant. But by far the majority of people who lived at Flowerdew Hundred and were responsible for leaving the record for the archaeologist to ponder were anonymous. Tenant farmers, native Americans, indentured servants, and slaves all have contributed to a story of everyday people doing commonplace things. In the acting out of these daily routines they were an integral part of early Chesapeake society, and their actions were guided by events of far greater magnitude taking place far beyond the boundaries of Flowerdew Hundred. Relatively few written records have survived for Prince George County, so archaeology has been the only way to recover these people, not only to restore them to their place in history but to account for an important part of both Virginia's and America's past that otherwise would have been forever lost.

Chapter Six

THE FIRST EXCAVATIONS at Flowerdew Hundred were undertaken when historical archaeology was still in its infancy. Few people were digging on historical sites, and only a handful of academic departments, mostly of anthropology, offered courses in the subject. As the field developed and changed, so did the questions that were being asked of the data, as well as the reasons for selecting this or that site to investigate. The archaeological research at Flowerdew had little or no direct effect on this development, at least in the earlier years, but Flowerdew is one of the few sites in America where there has been a continuous program of excavation since the early 1970s. For this reason, it is worthwhile to consider the archaeology of Flowerdew Hundred as it reflects the state of the field of historical archaeology today, and what considerations have brought it to this point.

An anonymous source has been quoted as saying that historical archaeology is an expensive way of finding out what we already know. A cynical assessment, perhaps, but unfortunately a significant amount of work has been done which has produced very little knowledge that is truly new, or that could not be obtained from documentary sources. Other scholars see the field as a handmaiden to history and would argue that archaeology has little to offer beyond a kind of footnoting of conventional historiography. This, too, has been the case in many instances. But neither position need be the proper one; it is less a matter of what is being done and more a matter of how one goes about doing it, and what questions are asked of the material. The complex set of sites and artifacts investigated at

Flowerdew Hundred, covering the entire period from the initial colonial settlement to the Civil War, serves to focus our attention on five issues that appear to have considerable relevance to historical archaeology and its future course of growth and development. Not everyone will agree with these statements; some of them are directly opposed to conventional wisdom, but they form a coherent set of related considerations.

The age of any site, in and of itself, is not a determinant of its significance.

The problem of defining the significance of a particular site is a complex one, and has yet to be fully resolved. What are the criteria that make this site an important one to excavate, and that one not? To be sure, older sites hold more promise for a number of reasons. The further one moves back in time, the less people are like their modern counterparts. It is mostly likely that if people were to meet one of the residents of Peirsey's Hundred, they would experience as great a degree of culture shock as they would upon meeting an exotic person alive today, from remote New Guinea or central Africa. This is for two reasons. First, almost everyone alive today participates to some degree in global culture. A memorable television show made of the people of Manus, Melanesia, first studied by Margaret Mead in the 1930s, showed us a woman, still dressed in traditional garb, listening to Beatles songs on a wind-up phonograph. Coca Cola is everywhere; it is not at all uncommon to see a Xhosa woman walking down the street in South Africa with anything from a single can to a case of Coke carefully balanced on her head. In these and a myriad other ways, the people of the world share today in whole segments of a common culture. Such is certainly not so in the case of seventeenth-century people. In no way do they share with us those things about our world with which we are familiar. These people were the carriers of a medieval tradition, with profoundly different attitudes, beliefs, and perceptions governing their every action. They would sound strange to the modern ear and have a different smell and a very different pattern of gesture and movement.

In the second place, we would be struck by the strangeness of these people because those of us who are European American and therefore their direct cultural descendants would expect them to be like us, simpler perhaps, but speaking the same language, having the same body of religious beliefs, and following the same general set of rules set forth in common law. But therein lies a genuine danger, for to project our values, ideas, and attitudes in a noncritical way on these seventeenth-century English folk could well result in a construction of their world fashioned along the lines of the one we inhabit today. From that it is an easy step to see them as earlier versions of ourselves and, in so doing, use our perceptions of the way they exemplify modern, late twentieth-century life as a justification for life as we now know it. This is the very important message conveyed by those archaeologists who adopt what is known as critical theory in their approach to their data. Much of modern critical theory is grounded in Marxist thought and sometimes tends to oversimplify matters by reducing every explanation to some aspect or another of the inequality between social classes. Inequality there has been, and certainly projecting present values on the past can serve to mask social differences by making it seem that those differences were always there—the "natural order of things." But there is more to it than that, and failure to recognize the complexity of the situation is always a real and present problem. In the final analysis we must always guard against making these seventeenth-century people simpler versions of ourselves. This is particularly true of the unlettered folks who made up the bulk of the population. Leading lives guided as much by folk beliefs as by those precepts handed down by the church and state, they were genuinely different in every way from anyone alive today, and these are the people that only archaeology can reach, for they wrote nothing down for later generations to read.

So, there is an intrinsic value in digging a site because it is quite early; we have access to a vanished world, one quite different from that we know today. In the Chesapeake, sites from the beginning years of the colony are not all that common. The first systematic archaeology on sites of this general period was carried out by the National Park Service at Jamestown, the first capital of Virginia. Jamestown seems to have been selected for this research both because it was the capital and because it was indeed an early settlement. Unfortunately, the earliest part of Jamestown had long since slipped into the river, so it became necessary to excavate other sites to get a picture of what life was like in the first decades of the seventeenth century. Half a dozen or so such early sites have been investigated, Flowerdew Hundred among them, and have provided the information to permit us to delineate in considerable detail what life was like at that time, how people constructed their houses, what they ate, what objects were commonplace in their homes, and how they laid out their settlements. All of this is most valuable, and more than justifies the effort and expense invested to make it known.

But age alone does not make a site worthy of attention. Our fixation on antiquity is probably a legacy of prehistoric archaeology, and the newspapers frequently carry articles on this or that person who has found this or that first or earliest thing at sites around the world. Great controversies develop over who has found the earliest example of Mayan monumental architecture in Guatemala or the first evidence of farming in the Near East. But such a fixation on primacy distracts from more fundamental issues of how all people in all of the past made their way through life. Process and change are the central concerns of any historical endeavor, whether historiographic or archaeological. There is every bit as much to learn from a slave cabin site of the early nineteenth century as from a well-documented early seventeenth-century settlement. The concept of information loss, first developed by Marley Brown, is relevant here. In assessing a site's significance, one important question must be asked. How much will we lose in information about its former occupants if we do not conduct excavations? If the amount is quite small, then the significance of the site is not all that great. On the other hand, the danger of losing a lot of knowledge by not excavating makes it more imperative that such a site be dug. What makes one site more liable to information loss than another is the degree to which it is represented in the documents. The less documentation, the more we stand to learn from excavation. Thus, site significance can be measured along two dimensions, time and documentary richness.

With this in mind, we can construct a simple chart which will allow us to look at the sites at Flowerdew Hundred, assessing their significance in terms of site-specific documentary and archaeological evidence (fig. 33). We can let the vertical axis represent time, from 1620 through 1820. This can be constructed to exact scale, with the earliest date at the bottom, the latest at the top. Unlike the vertical temporal scale, which can be precisely quantified, the horizontal axis represents relative degrees of documentary evidence for the sites in question. Assessing the value of documentary evidence is not subject to any kind of precision, but we can say that there is far more in the records dealing with the Willcox house than there is for either the slave cabin that stood nearby or even Miles Selden's house farther down the ridge. Thus, if the horizontal axis is divided with zero documentation on the left end and ample records on the right, the Willcox house will be to the right of the Selden house, which

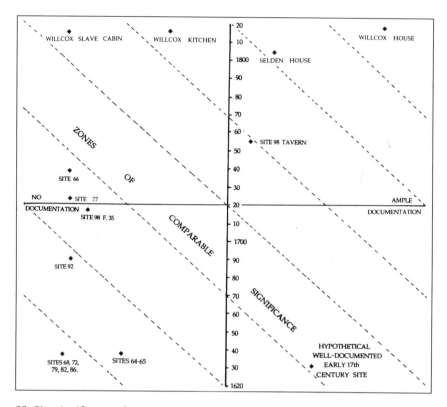

33. Site significance chart

in turn will be to the right of the slave cabin. The value of such a diagram becomes obvious when one proceeds to plot the various Flowerdew sites by the coordinates of documentation and time. It can be seen that such an exercise generates zones of comparability which run at a forty-five-degree angle to the two major axes. Sites within each zone are therefore roughly comparable in terms of significance, but this significance arises from the two criteria applied. Those sites of greatest significance are those which fall to the lower left-hand corner of the diagram, and those with the least significance are diagonally opposite at the top right-hand corner. All of this might seem to belabor the obvious, but the chart has value in the way it orders the sites relative to each other and the manner it presents quite clearly how both documentary richness and date can have relevance to a site's importance and how the Flowerdew sites relate to one another in this respect.

The little cluster of outlying sites that were contemporary with the Yeardley-Peirsey settlement thus has the greatest potential for producing new knowledge of all the sites on the plantation. The major settlement is only very slightly less significant because there are at least the two sets of census data, which in the case of the 1625 muster account not only for people but supplies, arms, livestock, and other equipment as well. Sites 66, 77, and 92 form a little cluster around the turn of the eighteenth century, as the second-rank set. All three have been excavated, and they produced important architectural data, insights into the emergence of early industrial activity on the plantation, and a large sample of Colono ware. Both the kitchen and slave cabin at the Willcox plantation form a third set, probably less significant than the earlier two but by far more so than the remaining three late eighteenth-century and early to mid-nineteenth-century sites. Within this last set, order of significance places site 98 first, the Selden house second, and finally the Willcox house third and least significant in terms of site-specific information loss.

This operation is a simplistic mechanical one and has been presented simply to demonstrate how we might look at these sites at Flowerdew in terms of significance and to make the all-important point that a later site with little or no documentation is every bit as important as an earlier one for which good records exist. A well-recorded early seventeenth-century site has not been found at Flowerdew, but such sites do exist elsewhere in the Chesapeake. Martin's Hundred has a richer body of written records than any Flowerdew site, and were it placed on this chart, it would fall somewhere in the zone containing sites 66, 77, and 92. If such a site was even more richly represented in the records, it would be in the same zone as the nineteenth-century Willcox slave cabin. Of course, implicit in the development of the chart is the premise that all other things would be equal, which we know in fact is not true. Until the sites were excavated, no one could have predicted the underground oven at site 82, the ice-house with its spectacular fill at Miles Selden's house, the roasting pits and other bloomery evidence at site 92, or the massive foundation of siltstone in the Yeardley-Peirsey settlement.

In a way the history of archaeology at Flowerdew Hundred roughly parallels the history of the plantation. The first several years of work were directed solely at the very early sites in the bottomlands, with the single exception of site 66. It was not until the 1980s that any excavations were carried out at sites from later times. In part this probably results from the earlier interest in old sites purely because of their age, a very common characteristic of historical archaeology in its earlier years. When the news of the stone foundation was first carried in *Time,* it was described as both the earliest stone house foundation in English America and the first example of a cruck house found on this side of the Atlantic.[1] There is nothing at all wrong about searching for the first this or the last that. We shall see that much of the value in such discoveries lies in their appeal to the emotions rather than to the intellect, an important point to bear in mind when thinking about presenting archaeology's findings to the public.

One should always use the archaeological record as a point of departure in conducting historical archaeological research.

Compared to the richness of the written record of at least some people in the past, the archaeological record is lean and impoverished. And while it is generally agreed that one should learn as much as possible about the site that is being investigated from the records, including researching titles to determine ownership, studying probate papers to get some idea of the worth of the estate of the site's former occupant(s), and consulting tax and census records to place the person(s) in question in the community, there is always the danger of the tail wagging the dog, in that much of the documentary research will have no real relationship to what is excavated. All too often, having done extensive documentary studies, the archaeologist will devote large sections of the report to historiography, often not all that well done, and fail to relate it in any meaningful way to the results of the excavations.

This approach has been called "unidirectional" and can be carried out in either of two ways.[2] On the one hand, the documents might tell us that the resident of the site we are excavating was quite affluent, and when the site is dug, confirming evidence of this appears. Having shown that the archaeology agrees with the tax and probate data, the research is considered finished. But in this case, the archaeology was hardly necessary, for the documents gave ample proof of what the digging only confirmed. The reverse of this approach takes the form of finding evidence in the ground that suggest the person who deposited it was of a certain social class and then confirming the archaeology with the documents after the fact. Both approaches operate on a one-way line of reasoning, which is either to confirm the documentary evidence with the archaeological or vice versa and then conclude that the job has been done. The literature of historical archaeology abounds with such reports, probably one reason that historical archaeology can be seen as "an expensive way of finding out what we already know."

The true relationship between the archaeological and written records is far more subtle, in terms of how it might be used to produce new insights into past time. In contrast to the unidirectional approach, what should be taken is a multidirectional one, using the material record as the point of departure. Since there is so much more in the documents, much of which has no reflection in the ground, to begin there risks following blind alleys; but if one begins with the archaeological records, all of it in some way was linked to that rich and active world to which the documents attest. *Multidirectional* in this context means working back and forth between the documents and what the site has produced, constantly refining and reformulating questions raised by one set of data by looking at it against the background of the other. Flowerdew Hundred gives a good example of how this approach works, and what can be gained from it. We have already discussed at length the patterns of pipe stem histograms and how they eventually led to some new and important insights into early Chesapeake society. Looking at this discussion in the light of how documents and artifacts can mutually complement each other, giving rise to new ways of thinking about the past, will make the multidirectional approach clear.

Beginning with a modest class of artifacts, smoking-pipe stem fragments, it was discovered that they suggested a three-phase pattern of settlement at Flowerdew Hundred. At this point the history of the Chesapeake in the seventeenth and eighteenth centuries was examined for possible causes of this pattern. The tobacco economy and slavery were tentatively suggested as somehow responsible for the three discrete groups of sites and the manner in which they occurred over time. The data from each set of sites were then examined to see if there was anything that would tend to support the suggested causes. Post-in-ground building was found in the first group of sites, more permanent construction and nascent industrial production was evidenced by the second group of sites, and the third group of sites, and only those, produced Colono ware. These associations directed our attention back to the historical record once again, for if the identification of Colono ware as largely the product of slave potters is correct, then what was known about Flowerdew's history and that of the Chesapeake in general directed our attention once again to the archaeological record. In this instance, the archaeology showed an absence of Colono ware during the long period of time when Africans were resident at Flowerdew. So, back to the documents once more, this time, Dell Upton's study of house size and servant-master relationships during the seventeenth century.[3] This study raised the distinct possibility that blacks resided in the same dwellings as Europeans and thus acquired a knowledge of European foodways and ceramic types. With this consideration in mind, the comparative archaeological record in Virginia and South Carolina made much greater sense, with the formal differences in Colono ware in each region arising from different patterns of residence and social interaction.

Two aspects of this study are of considerable importance and require further discussion. First, a significant part of it depended on another study which combined archaeological and historical data in a multidirectional way. The explanation of the relationship between earthfast building and tobacco monoculture was arrived at by having each body of data inform the other, raising new questions that could not have been formulated otherwise. The structural details and widespread use of earthfast structures could not have been known without the archaeology, but the relation between these structures and farming practices could not have been established without written records. Most important, the connection between impermanent construction and attitudes toward investment and its changes at different times in the region could not have been determined without combining both data sets to provide a more coherent and detailed aspect of early colonial life. Second, neither study combined site-specific documentary sources with site-specific archaeological materials. Rather, the site-specific archaeological data were considered against the background of the entire Chesapeake region. If forces and circumstances which affect an entire region were at work, these must have made their effect known at location after location.

Such use of regional historiography to attempt understandings of what may have happened at any single site is not as common in historical archaeology as it should be. The tendency usually is to gather a corpus of site-specific documentation concerning the site and its residents, to be connected to the archaeology in a unidirectional fashion. But there is a great danger in such site-specific research, for what may appear to be a particularistic event on one site may require broad comparisons to reach the proper explanation of what has been excavated. We have seen this to be true in the case of the filled icehouse at the Selden site, which, if our explanation is correct, needed an international perspective. One might even go so far as to say that site-specific documentary research is of little value in many cases, when we reach the level of more general, broader explanation. This is not to say that site-specific documentary materials are of no value per se, but rather that this value diminishes as we move to a more regional or national perspective. But there are aspects of site-specific documentary data that have value, as we have seen in the discussion of site significance. In the total absence of documents at a site—a slave cabin, for example—archaeology is the only source of site-specific information, and when we move to regional historiographies, sites such as this assume even greater significance. We will never know who lived at site 82 or any of the other small dwellings at Flowerdew, but through the use of the multidirectional approach, we can say some important things about how their lives were shaped and directed by events taking place far beyond their horizons.

If archaeology can do nothing else, when used properly it makes us ask different questions of the historical record, different by the very nature of the material being studied, which is rarely employed by those whose historiography is based on written sources. But archaeology holds the promise of more than just this, if we use it wisely and imaginatively. Such use depends heavily on maintaining the balance between two bodies of information which can support each other through the simple fact that they are complementary as well as supplementary, producing results that provide a more satisfactory explanation than would be forthcoming from either set of data alone. To be sure, the conclusions arrived at in the study of Colono ware and African-European interaction could have been arrived at by a route different from that taken, but regardless of the precise set of steps involved, it was necessary to incorporate artifacts in an archaeological context with documentary evidence to obtain the explanation. The pattern of distribution of Colono ware in time and space cannot be understood in the absence of documentary support. However, once this explanation has been provided, a dimension of black-white relations in seventeenth-century Virginia is clearer than it would have been if the archaeological data were not taken into account.

When the analysis of the Harrington pipe stem dates and settlement at Flowerdew was published, archaeologists at Colonial Williamsburg applied the same approach to a series of sites at Carter's Grove, the location of Martin's Hundred.[4] They reasoned that if the groupings of sites seen at Flowerdew Hundred were caused by regional changes, then other sites would show the same kinds of groupings. They pointed out the similarities between Martin's Hundred and Flowerdew: both started as fortified

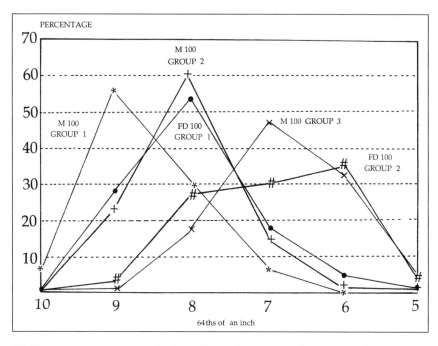

34. Pipe stem aggregate graphs from Flowerdew Hundred and Martin's Hundred

settlements, followed by a dispersal of the population over time, and both were consolidated into larger holdings from the mid-eighteenth century on. But there are also differences, most significantly the location of Martin's Hundred well down the river near Jamestown. When the pipe stems from eight sites were measured, and histograms constructed, once again three groups emerged (fig. 34). The first of these predates anything at Flowerdew. The second grouping makes an almost perfect match with Flowerdew Hundred's first set of sites, so close that the report calls it "actually unnerving." The third group at Carter's Grove matches the second Flowerdew Hundred grouping with a reasonable fit, and there is no fourth group, since sites from that time period were not examined. But the authors of the paper suggested that should such sites be studied, they almost certainly would conform to the third group at Flowerdew Hundred. The earlier group from Martin's Hundred in all likelihood represents settlement well before the 1622 uprising, as early as 1612, while settlement at Flowerdew only began in 1619. So it appears that the pattern of settlement history seen first at Flowerdew may have regional integrity, which comes as no surprise if the historical reasons for the groupings are what they seem to be. Once again we can see how a combination of site-specific archaeology and regional history produces new insights.

Historical archaeology is international in scope and must adopt an international comparative method to be of maximum value.

The similarities between plantation layouts in Ulster and Virginia, between vernacular building in English America and English South Africa, and the near identity seen in ceramic collections from British sites worldwide all serve to underscore the point made in the first chapter. Because historical archaeology deals with the spread of European culture into all parts of the world since the fifteenth century, we are provided with an opportunity to examine the way in which any European society, industrial or preindustrial, changed under different environmental constraints when implanted in a variety of new locations across the world. Dutch culture changed in different ways when established in New Amsterdam, Batavia, Brazil, and South Africa, as did British culture in Ulster, North America, Africa, and Australia; French culture in Acadia, West Africa, and the West Indies; and Iberian culture in the Americas. Examples already encountered at Flowerdew Hundred show the value of such a perspective, and it needn't be elaborated further. But there is another less obvious but equally important point that stems from this perspective, the matching of scale between one set of data with another and the way that affects the kinds of questions that are appropriate at different levels.

An example will make this clear. When the San Francisco Waste Water Project required excavation for sewer lines across the city, archaeology was carried out to salvage important materials that were encountered. In the course of the work, an extensive collection of later nineteenth-century Chinese material was recovered. The usual unavoidable constraints encountered in this kind of work made it impossible to assign precise locations to much of the material recovered. It was felt that as a result the value of the collection was seriously compromised. While this may be true in the context of research designs concerned with overseas Chinese in San Francisco, or a neighborhood of that city, it is emphatically not true in the context of research designs that are more global in scope. After all, the collection was recovered on the west coast of North America and, as such, can be compared with a similar collection from Tahiti, where Chinese merchants formed an important class in the nineteenth century, or with other Chinese settlements elsewhere in the world. Similarities and differences in this case are not considered in the context of status and occupation within a single community or in terms of the local economy at the time, but rather at the more coarse-grained but equally important level of international comparisons, such as a more global Chinese presence and its meaning in terms of the form that overseas Chinese culture assumes in all of those places where it is found.

As we widen the scope of our investigations, not only does the scale of provenience broaden, but the level rises at which artifact categories have meaning. Certainly the ratio between plain creamware and transfer-printed wares is important when we are concerned with economic dimensions of single households or communities. But at the level of international comparisons, ceramic categories that might seem hopelessly broad at the local level become important. Ratios between two major categories of ceramics, earthenware and porcelain, can reflect differences in international trade, access to certain markets, and different colonial situations. Archaeologists working on seventeenth- through early nineteenth-century sites along the eastern seaboard of the United States usually see porcelain as linked in some ill-defined way with status, or at least affluence. Rarely does porcelain exceed 20 percent of the ceramic sample, and sites yielding less than 5 percent are by far the more common. During the seventeenth century porcelain seems to have been quite rare at all economic levels. The entire porcelain sample from the Yeardley-Peirsey settlement can be held in two hands, and it is the largest collection of porcelain from this time period in all of the Chesapeake. Yet the owners of Flowerdew Hundred at the time were the two wealthiest men in the colony. Compare this pattern to those seen elsewhere. In the Indian Barracks at La Purisima Mission, California, porcelain comprised 24 percent of the total ceramic sample. In the mission dump, which presumably reflects the community as a whole, this number climbed to almost 40 percent. From largely unprovenienced collections salvaged during the construction of a shopping center in downtown Cape Town, South Africa, and from a late seventeenth- and early eighteenth-century fort in the same city, the percentage is even higher. Although precise figures are not available, examinations of these collections suggest that roughly half of the ceramics are porcelain, including a variety of coarse porcelains not found in North America.

What emerges from these comparisons are some very obvious questions, with equally obvious answers, but the simplicity of the situation should not detract from its importance. Southern California's position on the eastern rim of the Pacific basin must in some way account for the relative abundance of porcelain at La Purisima; one would predict similar quantities at other sites of the period along the entire Pacific coast, into South America. While it cannot be proved that the sixteenth-century porcelains recovered from Miwok Indian shell middens in Marin County, California, are evidence of Sir Francis Drake's presence there, they do attest to the presence of Manila galleons offshore with porcelain cargoes aboard. The high incidence of porcelain in the early Cape Colony is clearly the result of Cape Town's serving as a Dutch East Indies Company provisioning station between Batavia and Holland. The coarse porcelains, however, might owe their presence to a less-developed local ceramic industry and its products' superior durability over Dutch-produced earthenwares. Whatever the contributing factors, clearly ratios as simple as those between porcelains and earthenwares can inform us about the positions of different national colonial enclaves with respect to the rest of the world. Furthermore, these comparisons call into question an otherwise implicit assumption of the relationship between porcelain and status that may hold true in Anglo-America but could be just the reverse elsewhere.

The rise of the British ceramic industry to global dominance by the end of the eighteenth century provides us with yet another way of looking at ceramics on a worldwide scale. It is no surprise that virtually all of the ceramics found at Flowerdew Hundred are of English origin (German stonewares being the sole significant exception), and that they are almost identical to those found on sites from New England to the Deep South. But late creamwares and pearlwares are also found in South Africa, Polynesia, Russian sites in Alaska and California, and all of the California mission sites. Like Sherwin-Williams paint, they seem to cover the world. Given a common origin for these ceramics and a solid knowledge of their date of manufacture, we are provided with a degree of control which enables some very valuable comparisons. Since all of these ceramics have a common origin, differences between them in terms of function as indicated by shape and type of decoration, including choice of color, can better be explained in terms of the particular context from which they were recovered. We have already seen that there are significant color differences between African and American ceramics which share in a common origin, and that these suggest in turn certain differences in people's view of their world. Likewise, the prevalence of large transfer-printed pearlware bowls found on sites in the Anahulu Valley, Oahu, Hawaii,[5] probably results from their use as a replacement for the large wooden bowls used for poi. On the northwest coast of North America, similar bowls have long been the preferred containers used in the ritual exchange ceremony known as potlatch. In sum, having a specific body of well-dated ceramics which share in a common source allows comparisons on an international scale that offer new insights into the dynamics of colonial expansion and the impact that it had on local populations in many parts of the world.

Unlike prehistoric archaeology, historical archaeology has close connections with the humanities, particularly history and folklore.

By both its definition and name, historical archaeology incorporates the methods and perspectives of the historian as a part of its approach to the material world of the past. Ample examples of the value of historiography already have been encountered in the Flowerdew Hundred story. And, as we have noted, while the questions asked by historians of the archaeological record may be different from those posed by archaeologists trained in anthropology, they are no less important. In spite of this, historical archaeology has, since its emergence as a separate field of inquiry, been largely subsumed within the domain of anthropology. This relationship has tended to obscure some others equally important, not only with history but with folklore as well. The ties to history are reasonably obvious, but to folklore much less so. Why might this be so?

Back in the days when historical archaeology was barely visible on the American intellectual landscape, things seemed so simple and direct for those anthropologists who chose to become archaeologists. Archaeology, they were told, added a time dimension to the anthropologist's study of nonliterate peoples. How else were we to know the whole sweeping story of the development of human culture without going into the field with shovel and trowel, unearthing the evidence? And as social scientists, we properly valued the comparative cross-cultural approach taught us by archaeologists and ethnologists alike.

But all the while, a few people were doing something else at places like Colonial Williamsburg, Jamestown, Pecos Pueblo in New Mexico, and seventeenth-century farmsteads in New England and soon would begin excavations at Flowerdew Hundred. James Griffin's monumental *Archaeology of Eastern United States,* published in 1952, included an important essay by J. C. Harrington on the accomplishments and potential of historical archaeology. But the sites being studied had been occupied by people who also left us written records of their lives and those people could well have been the ancestors of the archaeologists who were studying them. Such research didn't quite fall into the model of cross-cultural comparative method, or if it did, the reasons were not recognized. Most of those doing this fieldwork were trained in departments of anthropology, and in some of their first encounters with historical sites, catastrophe resulted. Old southwestern hands were sent to New England to excavate sites of the Revolutionary War period, and prehistorians turned to historic sites in their specialty area with hardly any change in the field methods or research designs. But as the field expanded and matured, it became obvious that this kind of archaeology required a different body of knowledge and a different kind of training. An example will illustrate this quite clearly. It was common in the early years to classify ceramics as the prehistorians did, using descriptive categories derived from the pottery's physical attributes. Using this approach, shell-edged pearlware might simply be designated "blue on white" and, in at least one instance, was actually lumped with blue on white delftware, even though there is no relationship between the two types save a common color scheme. Classifications based on detailed knowledge of European ceramic history were called for, and in time this was realized and methods were changed accordingly.

While this recognition of the importance of history to the endeavors of historical archaeologists steadily became more obvious, it was not until the 1960s that another discipline was perceived as having close intellectual ties to historical archaeology. Perspectives offered by folklorists who specialize in folk life studies provide yet another way to consider the question of information loss. We have seen that the age of a site alone does not affect the amount of new information recovered. The degree of documentation is equally relevant, and this can lead to some interesting considerations, turning on the distinction between popular and traditional culture. Popular culture, depending in part on literacy and writing to contribute to its rapid dissemination and change, is more easily studied in the past without resorting to archaeological excavation. Traditional culture, the culture of the "folk," whoever they are exactly, is far less represented in the written record, and were we only to depend on documentary evidence, it would be virtually impossible to find out much about them in the earlier reaches of the past. But traditional culture changes with glacial speed and is very resistant to external influences. It is popular culture stood on it side, restricted in space rather than widespread, and changing slowly rather than rapidly. So it may be that since more traditional American culture has a higher potential for information loss, it should be among the primary concerns of historical archaeology.

Furthermore, since traditional culture changes so slowly, or at least did in the past, the possibilities of using contemporary data in understanding its past are significantly enhanced. There have been changes, to be sure. American folk culture along the eastern seaboard was altered significantly by changes wrought during the later years of the eighteenth century by major shifts in the economy, religion, and politics whose roots lay earlier in the Reformation and the opening of the New World. But along with change, there was continuity, and the modern differences between folk societies in America and that world of which most of us are a part impress all but the most insensitive of us. The Library of Congress's study of the Hammons family of eastern West Virginia, published as a two-record set and accompanying monograph, brings that difference alive in wonderful ways. Consider Sherman Hammons, narrating a childhood experience for folklorists in 1972. Hunting ginseng alone, having left his father while bringing back some stray sheep, he heard "something a-coming" down the river, making an awful noise. Looking up, he noticed that there were two of them about the size of hawks. As he tells it:

> And I heared that sound right there, and I said think to myself "What in the hell was that? What is that?" And it kindly made me feel better, and I said "Two devils a flying in the—"
>
> And I hit her. Now they'll—by George, I could run and was long winded. And I overtook 'em before they got to the—below Tea Creek there at the ford, at the crossing. And I said, "Tell me what in the Devil and Tom Walker was that?" I said. "Why," my dad said to me, said, "do you not know?"—ah, them was the first ones he ever saw, them was the first ones—he said, "Do you know what that was?" And I said, "No. Hell, I reckon it was two devils is all I know." "Why," he said, "them there wasn't devils, them was a airships," he said.[6]

One could understand a person in rural America in the early years of the century not having seen an airplane, but that he identified them as devils is the more remarkable thing about this tale. Other tales in the study are even more invested with a strong supernatural flavor, reflecting a view of reality both different from ours and probably very archaic.

While traditional culture is probably most susceptible to information loss in the absence of archaeology, its conservative nature allows careful comparison between modern expressive elements and the evidence of those in the ground, and it is so diverse regionally that a close accounting must be made of this diversity. Traditional culture is sufficiently different from the popular culture of which it is an enclave that it should be viewed from a perspective similar to that used by anthropologists in studying societies different from our own. Although few would argue that anthropological archaeologists should not attempt the study of traditional culture in the past, there is another group of scholars who have been working with such people for more years than we have spent studying their material remains in the archaeological record. Although folklorists are usually thought of as humanists who collect folk songs, riddles, jokes, and ghost stories, folklore has become increasingly concerned with the entirety of expressive traditional culture, including its material components. The study of folk life, long a tradition in Europe but not a significant part of folklore in America before the 1960s, entails many aspects of culture significant to the archaeologist. The creation of the American Folklife Center under public law 94-201 in 1977 is but one indication of the growing importance of folk life studies. As defined in the legislation creating the center, *American folk life* means: "Traditional expressive culture shared within the various groups in the United States; familial, ethnic, occupational, religious, regional. . . . Expressive culture includes a wide range of creative and symbolic forms, such as custom, belief, technical skill, language, literature, art, architecture, music, play, dance, drama, ritual pageantry, and handicraft. Generally these expressions are learned orally . . . and perpetuated without formal instruction or institutional direction."

Of course some aspects of expressive culture are beyond archaeological recovery, but since folk life studies embrace all of those aspects of life which do leave physical remains—architecture, technical skill, handicraft—archaeology holds the promise of providing a time depth to the work of folk life specialists. In a way, they have been working on the very surface of the cultural landscape beneath which so much of our attention has been directed. When we bring traditional culture into the present, we are far more likely to meet folklorists than ethnographers, humanists rather than social scientists. That such was not so obvious until recently is only natural, since archaeologists have only lately advanced into more recent periods within the last five centuries of the American past, and folk life studies are now enjoying a greater emphasis within folklore. And since one could only learn archaeology in an anthropology department, programs in folklore that emphasize folk life could not offer that research technique to their students.

So there are no compelling reasons for historical archaeology to remain the exclusive property of anthropologists. This is not to suggest that historical archaeology should become a part of folk life studies exclusively, but that a recognition of the affinity between the two fields might be beneficial. There are a variety of archaeological problems that do not fall within the confines of traditional culture—urban archaeology, the archaeology of American popular culture, military archaeology, and many others—but we should recognize more clearly the close affinity between a kind of archaeology wherein information loss is likely to be great and the study of traditional cultures by people well trained to do so. Folklorists have produced important studies which call attention to the relationship between the material world and the thought and behavior which gives it shape. Henry Glassie's study of folk housing in middle Virginia, John Vlach's research on African-American material culture, Robert Saint George's work on furniture and architecture in New England, and Jay Anderson's studies of foodways are but a few examples.

It can be honestly said that the twenty-five years of archaeology at Flowerdew Hundred have been informed in a major way by the work of both historians and folklorists. Although all of the excavations have been directed by archaeologists trained in anthropology departments, the scholarship and insights of historians such as William Kelso, Cary Carson, and Edmund Morgan and those trained in folk life studies, Henry Glassie, Robert Saint George, Dell Upton, and John Vlach, to name but a few, have combined to create a coherent account of the history of the plantation which would have been impossible to construct without all of their contributions.

The tangibles of historical archaeology should appeal both to the emotions and the intellect.

The artifacts, house foundations, abandoned wells, and all of the other things unearthed by the historical archaeologist provide us with a rich basis on which to construct varied aspects of the past, often detailing that which otherwise might remain less visible in a very literal sense. As such, they provide access to new perspectives on history and, in many instances, permit us to make statements that would have been impossible in the absence of the archaeology. This is the very justification for the field of historical archaeology and for the time and money expended to pursue it. Done with imagination and sophistication, historical archaeology can be a challenging and stimulating intellectual endeavor. But there is yet another aspect of all the things uncovered by trowel and shovel, one that is less recognized and thus less valued. Objects of the past, when encountered in the course of excavation, can have a powerful emotional impact. When an excavator stands in the bottom of a cellar that was filled in 1675, that person is the first to have stood there in over three centuries. Past and present come face to face. The same applies to all of the material remains of the past. When the complete, unused clay smoking pipe was taken from the fill of the grave at site 77, the person lifting it from the soil was the first to touch it since it was placed there at the end of the eighteenth century. Thus an object can link, in a subtle but powerful way, two human beings separated by hundreds of years. We cannot say the same about objects of equivalent age that have somehow survived above ground, for they have been handled by hundreds, perhaps thousands of people during the time since they were originally in use. But with archaeological objects, which have been sleeping in the earth since they were buried, one senses the connection with their owners in a much more direct way, one that, if any emotion can be expressed, is a moving experience.

In the cellar fill at site 77, a gold wedding band was discovered, inscribed on its inner side "Grace mee with Acceptance" and the initials "T.S." While we will probably never know who "T.S." was—the initials don't match with those of Flowerdew Hundred's late seventeenth-century residents—there is something particularly poignant and intimate about this ring. Of solid gold, it was an object of great value and probably held strong sentimental associations with its owner. How it made its way into the fill of the cellar is a mystery, but one can be almost certain that its loss was mourned. Lesser objects also can generate feelings of a very personal and emotional nature about their place in a vanished world and what they meant to their owners. So we must not deny the emotional content of these things, but rather use it to communicate to others about life in a far-off time.

It is when the emotional content of things archaeological becomes confused with the intellectual that problems arise. One can look at the ironworking equipment excavated at seventeenth-century Saugus, Massachusetts, and point out that it tells us a story of early attempts at ironworking in the colonies. One can even suggest that the excavation led investigators into an area of research on seventeenth-century ironworking technology that might not have been pursued had the archaeology not been done. Yet the equipment per se tells us little new about the subject, for its history and technology are well attested to in documents, both in England and America. So to justify the excavation of this material in intellectual terms is a shaky proposition. But the emotional impact of these objects is palpable, reminding us in ways that no written account could of what it must have been like on the rough New England frontier, trying to develop a technology in the face of considerable odds. That is the true value of such objects, and if by some chance they do contribute something new to our knowledge, well and good. But they need not, for it is their intrinsic worth that justifies their recovery.

It is this worth that many archaeologists have slighted in their writing and communicating of their findings to the lay public. Only specialists are interested in the things usually written in a site report. What person, other than an archaeologist, or perhaps an historian, really cares about the relative price of various types of pearlware in the early nineteenth century? The knowledge is important, to be sure, but other things need communicating as well. For if we as archaeologists are to continue our work, it must be in the context of public understanding and support. Fortunately this is taking place, though not as much as one would like. So-called popular writing is looked down upon by many professionals, who somehow think that for a work to be important it must somehow be painfully technical. But this need not be so, for the same information can be communicated in a lively way, using simple, declarative English. Books such as Noël Hume's *Martin's Hundred* or William Kelso's *Kingsmill Planta-tion*[7] attest to this fact. Museums also can play a significant role in communicating such understandings. The new museum at Martin's Hundred and the Museum of the City of London are but two excellent examples of how the complexities of historical archaeology can be presented in an engaging and powerful way, to catch the viewers' interest and teach them something as well. It was Charles P. Wilcomb, an early museum worker, who said it best in the first years of the twentieth century, that what we do is education in the guise of entertainment.

If the field is to continue to grow and prosper and contribute to our fund of genuine knowledge about the past, it is imperative that the emotional content of our subject not be buried beneath layers of dull, jargon-ridden reporting. This is not to suggest that the artifacts themselves be celebrated at the expense of the all-important contextual dimension of the data. New understandings and constructions of the past can only come from the various associations and quantifications that we extract from our material; context is all important. But we must also not lose sight of the intrinsic worth of those things which we also consider in their associational context and relative numbers. The Flowerdew Hundred story provides ample demonstration of this fact. The small fragment of Alice Thorowden's clay pipe is both evocative as a single object and informative as but one of hundreds of similar fragments, which taken together allow us to look into the past with a clearer, sharper vision.

July 27, 1983, Flowerdew Hundred Farm, Virginia

Alicia Trow wakes to her alarm just before daybreak. She has been here in Virginia now for ten days, working with a group of University of California students who have been excavating an early seventeenth-century site on the banks of the James River. The work has been tiring at times, up at the crack of dawn, on the site by 6 A.M., long hours spent under the relentless sun, but it has been exciting as well. One never knows what will be uncovered next, and already she is struck by how very different all these fragments of the past are from what she knows from her own experience. Alicia grew up in Los Angeles and has never before traveled in the United States east of Las Vegas, although she has made two trips to Europe. So it was with a real sense of expectation, tempered with slight apprehension, that she left for Virginia. To her surprise, it is in many ways more exotic than anything she has experienced in Europe. The local folks speak a strong dialect; in fact, some of the farm workers she can hardly understand. Fireflies are a new source of wonder, and she has never experienced thunderstorms such as have swept through the field camp on two occasions, ripping tents, setting trees afire, and lighting the horizon brighter than day. The choruses of frogs at night echo across the pond below the camp, and the cry of whipoorwills can be heard in the forest across the fields. On some mornings layers of mist float just over the corn, to vanish when the sun appears above the trees to the east, red as blood.

Alicia leaves her tent and joins her fellow workers around the coffeepot in the kitchen area. Quiet greetings are exchanged; people are still waking up. At 6 A.M. everyone climbs into a brown pickup, and it drives off, down along the dusty farm road, between tall rows of corn, throwing up a plume of dust that slowly settles in its wake. A mother wild turkey and her chicks run in front for a few hundred feet and then veer into the corn. As the truck turns left, along the river, two adult bald eagles rise from the trees and fly past, so close that their yellow eyes can be seen against the white of their heads.

Arriving at the site, the crew removes the plastic sheeting covering the excavated area. Just to the west the outlines of a massive stone foundation can be seen beneath other plastic covering. Alicia's task today is to begin the removal of fill from a small trash pit that she had defined yesterday, an oval stain, a foot or so across. Taking her trowel and carefully scraping the soil, she begins to remove the contents of the feature, looking closely for whatever objects might be in the fill. She is rewarded immediately. Bits of blue and white pottery appear, then damp brown fragments of bone. Several twisted bits of rusted iron come to light; these will have to be cleaned carefully before anyone can say just what they might be. Then, common pins, parts of a square bottle of light green glass, a bit of brown pottery with a curious face looking back at Alicia, more blue and white pieces, twisted strips of lead—the pit is a rich deposit, and she removes everything with great care and places each object in a plastic bag which has been labeled. Three hours later she is nearing the bottom, and there, lying close together, are a number of smoking-pipe fragments, obviously enough to make a whole pipe when they are joined back together. Alicia notes that the bore of the stem fragments is quite large; she guesses $8/64$ inch. While this comes as no surprise, since the site is an early one, Alicia lets her imagination go for a few moments. Whose pipe was this? Why are all of the pieces here in one small area? Who might have smoked it last, and what were the person's thoughts about this very place where she was kneeling, carefully troweling the last of the dark soil? Alicia, meet Alice. The circle is closed.

Appendix
Notes
Index

Ceramics from Flowerdew Hundred

R EADERS OF REPORTS AND SUMMARIES of the findings of historical archaeology are often struck by what seems at times to be an undue emphasis on the various bits of pottery recovered from sites. But in fact, the analysis of historical ceramics can provide us with significant insights, as our examination of Flowerdew Hundred has shown. The importance of ceramic studies in piecing together a coherent story of the past results from three factors, the fragility, durability, and universality of pottery in the world of early America. Pottery breaks relatively easily; the use life of any commonplace ceramic object is of the order of five years, and often even less. There are some exceptions, particularly in the case of items such as porcelain tea services, which had a high value and were cared for to a greater degree than everyday, simple utility items. But since most ceramic pieces in a household would have passed into the refuse heap in a relatively short period of time, we can be confident that any given site was occupied sometime during the period when particular types of pottery were being manufactured. Since such dates are based on a number of different types, each with a different production period, it is often possible to assign a reasonably precise date for the occupation of the site based on the ceramic evidence alone. It is certainly true that of any type of artifact recovered in large quantities, ceramics provide us with the most dependable *termini post quem* since we know the production dates for most European ceramics. It is the beginning date that is relevant; if the manufacture of creamware did not begin until 1765, then any site or feature producing this type of pottery must have a *terminus post quem* of that date. Paradoxically, although whole ceramic vessels are fragile, their fragments are exceptionally durable once they have become buried. A fragment of German stoneware looks as if it is part of a vessel made only yesterday. This excellent state of preservation ensures that ceramic samples will not be skewed by differential survival. What is recovered is a reasonable representation of what was in use in the household when the site was occupied. The universal use of ceramics, for the most part as a central component of food preparation and consumption, provides the assurance that differences between sites can be measured and interpreted along a relative scale, rather than on a presence/absence basis. Every household owned some pottery; every site produces its fragments.

Ceramics from historical sites, Flowerdew included, fall into three broad categories, earthenware, stoneware, and porcelain. Earthenware is fired at a relatively low temperature, on the order of 800° Celsius, producing a body that is permeable by water. For this reason, glaze must be applied to render the vessel impermeable. The commonest glaze used on earthenware is known as lead glaze, made from either lead sulfide or lead oxide. When applied to a ceramic piece and fired, the lead compound fuses with silica in the clay to form a glassy coating. A pure lead glaze is colorless, but impurities or additives can lend color—most commonly, yellow from iron, green from copper, and purple from manganese. Whether colored or clear, lead glazes are transparent, and the underlying body, and decoration on it, can be seen through the glaze.

A variety of decorative techniques were used on earthenware. Slipwares were produced by applying a white slip, a solution of ball clay the consistency of heavy cream, to a red body in a variety of ways. Examples of such slipwares illustrated here include North Devonshire sgrafitto ware, Astbury-type ware, agate ware, and combed and dotted slipware, all in browns and yellows. Later earthenwares have white bodies and are decorated by painting, dipping in slip, glazing with a mixture of colors to produce a clouded effect, and transfer printing, both underglaze and overglaze. Pieces illustrated here of this later white-bodied ware include Whieldon clouded ware, creamware, and pearlware. The addition of tin oxide to a lead glaze produces a slightly thicker opaque white glaze, on which painted decoration can be applied. In this case, the body beneath is covered with the opaque glaze and is not seen through it. Such wares, collectively known as tin-glazed or tin-enameled wares, were produced on the Continent and in Britain from the fifteenth century on and are known variously, depending on where they were made, as majolica (Spain, Portugal, Holland, and Italy), faience (France), and, most familiar, delftware, made in both Holland and England. Most delftware found on American sites dating from the second half of the seventeenth century and later is of British manufacture, although Dutch examples also occur, particularly during the first half of that century. It is conventional to designate tin-glazed wares from Holland as Delft ware and those from England as delftware. Tin-glazed wares are very common on Flowerdew Hundred sites. The example illustrated here is typical of eighteenth-century delftware.

Stonewares are fired at a higher temperature, on the order of 1,400° Celsius, and are thus impermeable to water. They were glazed, however, to make them more attractive and easier to keep clean. Because a lead glaze will not withstand the high firing temperatures, a salt glaze was employed, achieved by adding sodium chloride to the kiln as the pottery was fired. The salt vaporizes and condenses on the body of the vessel, producing a shiny glaze with slight irregularities which resembles an orange peel. Stonewares from American sites were made both in the Rhineland and in Britain. The earlier ones are of German origin and include three different types, illustrated here. Brown stoneware bottles, the so-called bellarmines, are found on every Flowerdew Hundred site of the seventeenth century. Equally common, and lasting throughout the seventeenth century and through the third quarter of the eighteenth, are blue on gray stonewares produced in the Westerwald district. These sometimes were additionally colored with manganese after ca. 1660. Eighteenth-century examples frequently carry the initials of the various English monarchs, including AR, for Queen Anne, and GR for the three Georges. These were made expressly for the British market. A third type of German stoneware is a monochromatic brownish gray and was made in Höhr. Stoneware production began in Britain in 1671, and the earlier examples are brown on gray pieces, commonly mugs. Brown stonewares were made in England well into the eighteenth century, but beginning in the first quarter of that century, white stonewares joined them. The earlier type, made white by dipping the gray body in white slip and putting a band of iron oxide around the lips and on the handles of mugs, is known as slip-dipped stoneware. This type of stoneware is represented on American sites almost exclusively in the form of mugs, as illustrated here. By the 1730s a more refined white stoneware was being produced in the Staffordshire factories, not requiring a slip because the clay of the body was white rather than gray. Examples illustrated here include a molded plate, a type known as scratch blue—decorated by incising the body and filling the lines with blue pigment—and so-called debased scratch blue, commonly found in chamber pots and jugs with GR initials on applied medallions.

Porcelain is a ceramic fired at the same high temperature, but unlike stoneware, it is white and translucent. It has a vitreous body, made from a mixture of clays, the most critical ingredient being kaolin. Porcelain is a Chinese invention, and virtually all porcelain recovered from Flowerdew Hundred sites is of Chinese origin, although a European porcelain industry was developed by the mid-eighteenth century. These porcelains are either decorated in blue painted designs underglaze or polychrome enamel overglaze. The pieces illustrated here are typical mid-eighteenth-century Chinese blue on white.

Unless otherwise noted, the pieces illustrated here were carefully selected from the extensive collections of Colonial Williamsburg Foundation to represent actual vessels in use at various Flowerdew Hundred sites as evidenced by recovered fragments. These whole objects convey a better sense of ceramic diversity than the fragments would. Every effort was made to select vessels that matched pieces from Flowerdew Hundred sites, and in all cases the match is very close, if not exact. The type name is given in each instance, along with place of manufacture, color, form (such as plate, bowl, or mug) as they have been found at Flowerdew Hundred, the sites mentioned in the text which produced them, and the approximate dates of production. These dates are not absolutely precise, for in many instances there is not complete agreement on the subject, but they can be taken as probably correct within a quarter of a century. This corpus of ceramic material is not in any way comprehensive but rather illustrates those types which are more frequently encountered in excavation or are particularly significant for other reasons. The primary reason for presenting this material is to provide the reader with some sense of the richness of the ceramic materials recovered from sites at Flowerdew Hundred.

I.

North Devonshire
sgrafitto ware

England

1630–1700

Brown incised designs
on yellow

Plates

Site 72

Photograph: a modern
reproduction in the
author's possession

2.

Combed and dotted
slipware

England

1680–1780

Brown on yellow

Two-handled cups,
plates

Sites 66, 77, 92, 98

All photos unless
otherwise indicated
are courtesy of
Colonial Williamsburg
Foundation

3.

Astbury-type ware

England

1725–50

Pale yellow on brown

Bowls

Site 98

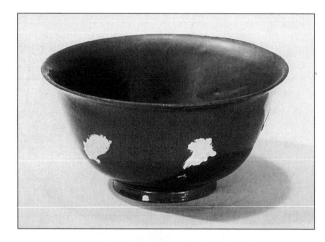

4.

Agate ware

England

1740–75

Yellow on brown,
agatized body of red
and yellow clay,
shows through glaze

Bowls, plates

Site 98

5.

Jackfield-type ware

England

1740–90

Shiny black

Tea sets

Sites 98, 113

6.

Whieldon clouded ware

England

1750–75

Various combinations of brown, purple, yellow, and green

Plates

Sites 98, 113

7.

Creamware

England

1765–1830

Cream colored,
polychrome painted,
red and black transfer
printed, plain

Plates, bowls, mugs

Sites 98, 113

8.

Pearlware, shell edge

England

1780–1830

Green or blue on
white

Plates

Sites 98, 113, 114

9.

Pearlware, painted

England

1780–1820

Blue on white

Bowls, pitchers

Sites 98, 113

10.

Pearlware,
polychrome painted

England

1790–1820

Polychrome on white,
floral motifs

Tea services, pitchers

Sites 98, 113

11.

Pearlware, annular

England

1790–1830

Various colors in
concentric bands

Bowls, mugs, pitchers

Sites 98, 113, 114

12.

Pearlware, transfer
printed

England

1790–1830

Blue, black, green,
pink, red

Sites 98, 113, 114

(Illustrated piece from
Davenport Factory, ca.
1810, original "willow
pattern")

Private collection

13.

Delftware

England

1570–1770

Blue and white,
polychrome

Plates, bowls,
ointment jars

Sites 64, 65, 66, 72,
77, 92, 98, 113

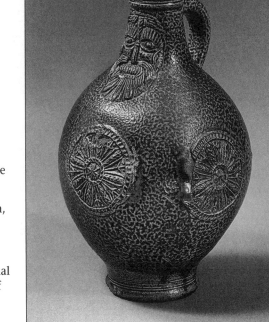

14.

German brown
stoneware, bellarmine
jar

Cologne and Frechen,
Germany

1550–1700

Brown with occasional
splashes of blue, buff

Sites 64, 65, 72, 77,
82, 113

15–16.

Westerwald stoneware

Germany

1590–1780

Blue on gray, often
with purple as well

Jugs, mugs

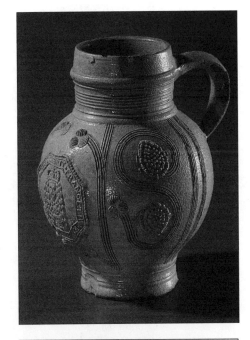

17.

Höhr-type stoneware

Germany

1690–1714

Gray

Jugs

Sites 66, 98

18.

Brown stoneware

England

1671–1775

Brown

Mugs, cups, jars

Sites 66, 77, 92, 98,
113

19.

White slip-dipped salt-glazed stoneware

England

1720–70

White with brown band around rim

Mugs

Sites 77, 98

20.

White salt-glazed stoneware

England

1730–80

White

Plates, bowls, mugs, tea sets

Sites 98, 113

21.

Scratch blue
stoneware

England

1745–75

Blue and white

Bowls

Site 98

22.

Debased scratch blue
stoneware

England

1765–90

Blue and light gray

Jugs, chamber pots

Site 98

23.

Chinese export porcelain

China

1675–present

Blue and white, polychrome painted

Tea services

Sites 98, 113, 114

Notes

Chapter One

1. J. C. Harrington, "Dating Stem Fragments of Seventeenth and Eighteenth Century Clay Tobacco Pipes," Archaeological Society of Virginia, *Quarterly Bulletin* 9, no. 1 (1954): 10–14.
Harrington's discovery was not an instantaneous one, but rather emerged slowly over a considerable period of time. As Pinky describes it in an Aug. 3, 1992, letter to the author: "You ask when it occurred to me that stem bores varied, and that the earlier ones (based on bowls) are smaller. This all came gradually over several years, and not as a sudden revelation. . . . after we moved to Richmond in 1946 I finally got around to exploring this tentative observation. So I brought all the pipe bowls in the Jamestown collection to Richmond, and continued to gaze at the stem bores, finally deciding that there was something here that needed further analysis. My problem was how to quantify this; that is, how to measure the bore sizes. I couldn't just say 'little,' 'medium,' 'big,' etc. As we always do when faced with a problem, Virginia and I were discussing this over a martini, or two, and she said: 'Why not use those steel drills you have down on your work bench?'
"And thus it all began some time in 1952 or '53. I then spent evenings measuring stem bores [and] worked up my paper, but at that time there was no broad outlet for articles on historical archaeology, so I gave it to the Archaeological Society of Virginia. The rest is 'history.'"
2. James Deetz, *In Small Things Forgotten: The Archaeology of Early American Life* (Garden City, N.Y.: Anchor Press, 1977), p. 5.
3. Ibid., pp. 29–30.
4. Ibid., p. 24.
5. Ivor Noël Hume, *Historical Archaeology* (New York: Norton, 1975), p. 3.
6. Henry Glassie, *Folk Housing in Middle Virginia: A Structural Analysis of Historic Artifacts* (Knoxville: Univ. of Tennessee Press, 1975), p. 12.
7. Jack Larkin, *The Reshaping of Everyday Life, 1790–1840* (New York: Harper & Row, 1988), p. xiii.

Chapter Two

1. Fraser D. Neiman, *The Manor House before Stratford* (Stratford, Va.: Robert E. Lee Memorial Association, 1980).
2. Quoted from Virginia M. Meyer and John Frederick Dorman, *Adventures of Purse and Person: Virginia, 1607–1624/5*, 3d ed. (Richmond: Dietz Press, 1987), pp. 22–24. See ibid., p. xxiii, for the dating of the muster, which was taken on January 20, 1624 by the Julian calendar in use then and 1625 by the Georgian calendar we use today.
3. Quoted from H. R. McIlwaine, ed., *Minutes of the Council and General Court of Colonial Virginia, 1622–1632, 1670–1676* (Richmond: Virginia State Library, 1924), p. 120.

4. Quoted from Edward D. Neill, *Virginia Carolorum* (Albany, N.Y., 1886), p. 404.

5. Jeffrey P. Brian et al., *Clues to America's Past* (Washington, D.C.: National Geographic Society, 1976), pp. 140–41.

6. Douglas Ubelaker, "Three Early Colonial Skeletons from Site PG3, Virginia," MS on file, Flowerdew Hundred Foundation.

7. Ivor Noël Hume, *Martin's Hundred*, rev. ed. (Charlottesville: Univ. Press of Virginia, 1991).

8. Ibid., pp. 53–59.

9. Anthony Garvan, *Architecture and Town Planning in Colonial Connecticut* (New Haven: Yale Univ. Press, 1951).

10. "Lists of the Livinge & Dead in Virginia Feb. 16th, 1623 [1624]," in *Colonial Records of Virginia* (Baltimore: Genealogical Publishing Co., 1964), pp. 40, 63.

11. Ibid., p. 9.

12. Quoted in Edmund S. Morgan, *American Slavery, American Freedom: The Ordeal of Colonial Virginia* (New York: Norton, 1975), p. 122.

13. Ibid., p. 123.

14. Ibid., p. 120.

15. Cary Carson et al., "Impermanent Architecture in the Southern American Colonies," *Winterthur Portfolio* 16 (1981): 135–96.

16. Morgan, *American Slavery, American Freedom*, p. 112.

Chapter Three

1. Eve Gregory and Pat McClenny, "Flowerdew Hundred: A Brief History," MS on file, Flowerdew Hundred Foundation.

2. The Southside part of Charles City County became Prince George County in 1703.

3. Ann B. Markell, "44PG92: Flowerdew Hundred Site Report," January 1990, MS on file, Flowerdew Hundred Foundation.

4. Samuel Symonds to John Winthrop the Younger, 1638, *Collections of the Massachusetts Historical Society*, 4th ser., 7 (1865): 118–20.

5. For a full discussion of the evidence for bloomeries at Flowerdew Hundred, see Ann B. Markell, "Manufacturing Identity: Material Culture and Social Change in Seventeenth Century Virginia," (Ph.D. diss., Univ. of California, Berkeley, 1990).

6. Fred Kniffen, "Louisiana House Types," Association of American Geographers, *Annals* 26 (1936): 177–93.

7. William Bradford, *History of Plymouth Plantation, 1620–1647*, 2 vols. (Cambridge, Mass.: Massachusetts Historical Society, 1912), 2:214–15.

8. William Kelso, "Big Things Remembered: Anglo-Virginia Houses, Armorial Devices, and the Impact of Common Sense," in Anne Elizabeth Yentsch and Mary C. Beaudry, eds., *The Art and Mystery of Historical Archaeology: Essays in Honor of James Deetz* (Boca Raton, Fla.: CRC Press, 1992), pp. 195–213.

9. Dell Upton, "The Origins of Chesapeake Architecture," in *Three Centuries of Maryland Architecture* (Annapolis: Maryland Historical Trust, 1982), pp. 44–57.

10. Ibid., p. 48.

Chapter Four

1. Ivor Noël Hume, "An Indian Ware of the Colonial Period," Archaeological Society of Virginia, *Quarterly Bulletin* 17, no. 1 (1962): 1–12.

2. Ibid., p. 4.

3. Frank G. Speck, *Chapters on the Ethnology of the Powhatan Tribes of Virginia,* Indian Notes and Monographs 1, no. 5 (New York: Heye Foundation, 1928).

4. Lewis Binford, "Colonial Period Ceramics of the Nottoway and Weanock Indians of South-eastern Virginia," Archaeological Society of Virginia, *Quarterly Bulletin* 19, no. 4 (1965):78–87.

5. Leland Ferguson, "Looking for the 'Afro' in Colono-Indian Pottery," in Robert L. Schuyler, ed., *Archaeological Perspectives on Ethnicity in America* (Farmingdale, N.Y.: Baywood, 1980), pp. 14–28.

6. Thomas Jefferson, *Notes on the State of Virginia,* ed. William Peden (Chapel Hill: Univ. of North Carolina Press, 1955), pp. 96–97.

7. Richard R. Polhemus, *Archaeological Investigations of the Tellico Blockhouse Site (40MR50): A Federal Military and Trade Complex,* (report submitted to the Tennessee Valley Authority, 1977), published as TVA Reports in Anthropology, no. 16, p. 258.

8. Upton, "Origins of Chesapeake Architecture."

9. Morgan, *American Slavery, American Freedom,* p. 155.

10. Leland Ferguson, *Uncommon Ground: Archaeology and Early African America, 1650–1800* (Washington, D.C.: Smithsonian Institution Press, 1992), pp. 96–107.

11. John Michael Vlach, *The Afro-American Tradition in Decorative Arts* (Cleveland: Cleveland Museum of Art, 1978).

12. Susan L. Henry, "Terra-Cotta Tobacco Pipes in 17th Century Maryland and Virginia: A Preliminary Study," *Historical Archaeology* 13 (1979): 14–37.

13. Ibid., p. 14.

14. Matthew Charles Emerson, "Decorated Clay Tobacco Pipes from the Chesapeake" (Ph.D. diss., Univ. of California, Berkeley, 1988).

15. Henry, "Terra-Cotta Tobacco Pipes in 17th Century Maryland and Virginia," p. 14.

16. Quoted in Mary Beaudry, "Wilkin's Ferry: A Late Eighteenth-Century Site in Prince George County, Virginia," MS, Department of Anthropology, College of William and Mary, Williamsburg, Va., 1980, p. 12.

17. Geoff. Egan, Susan D. Hanna, and Barry Knight, "Marks on Milled Window Leads," *Post-Medieval Archaeology* 20 (1986):303–9.

18. Ibid., pp. 303–4; Noël Hume, *Martin's Hundred,* p. 324.

19. Egan, Hanna, and Knight, "Marks on Milled Window Leads," p. 307.

Chapter Five

1. Margaret Eileen Scully, "Miles Selden of Flowerdew Hundred, Virginia: Defrosting the Ice House and Tidying Up the Cellar. Is His Garbage a 'Sign of the Times'?" (Ph.D. diss., Univ. of California, Berkeley, 1986).

2. Geoffrey Moran, Edward Zimmer, and Anne E. Yentsch, *Archaeological Investigations of the Narbonne House, Salem Maritime National Historical Site,* Cultural Resources Management Study no. 6 (Boston: Division of Cultural Resources, North Atlantic Regional Office, National Park Service, U.S. Department of the Interior, 1982).

3. Diana Edwards, Steven R. Pendery, and Aileen Button Agnew, "Generations of Trash: Ceramics from the Hart-Shortridge House, 1769–1860," *American Ceramic Circle Journal* 6 (1988): 29–52.

4. Vernon G. Baker, "Archaeological Visibility of Afro-American Culture: An Example from Black Lucy's Garden," in Schuyler, *Archaeological Perspectives on Ethnicity,* pp. 29–37.

5. Margot Winer and James Deetz, "The Transformation of British Culture in

the Eastern Cape, 1820–1860," *Social Dynamics* 16, no. 1 (University of Cape Town, African Studies Department, 1990): 55–75.

6. Patricia E. Scott and James Deetz, "Building, Furnishings, and Social Change in Early Victorian Grahamstown," ibid., pp. 86–89.

7. John Solomon Otto, "Race and Class on Antebellum Plantations," in Schuyler, *Archaeological Perspectives on Ethnicity,* pp. 3–13.

8. Lawrence William McKee, "Plantation Food Supply in Nineteenth-Century Tidewater Virginia" (Ph.D. diss., Univ. of California, Berkeley, 1988).

9. Gene Prince, "Photography for Discovery and Scale by Superimposing Old Photographs on the Present-Day Scene," *Antiquity* 62 (1988): 112–16.

Chapter Six

1. "Treasures of Flowerdew," *Time,* Nov. 20, 1972, pp. 66–67.

2. James Deetz, "American Historical Archaeology: Methods and Results," *Science,* Jan. 22, 1988, pp. 362–67.

3. Upton, "Origins of Chesapeake Architecture."

4. Andrew C. Edwards and Marley R. Brown III, "Seventeenth-Century Settlement Patterns: A Current Perspective from Tidewater Virginia," in Theodore R. Reinhart and Dennis J. Pogue, eds., *The Archaeology of 17th-Century Virginia* (Richmond: Archeological Society of Virginia, 1993), pp. 285–309.

5. Patrick V. Kirch and Marshall Sahlins, *Anahulu: The Anthropology of History in the Kingdom of Hawaii,* vol. 2, *The Archaeology of History* (Chicago: Univ. of Chicago Press, 1992), p. 182.

6. Carl Fleischhauer and Alan Jabbour, eds., *The Hammons Family: A Study of a West Virginia Family's Traditions* (Washington, D.C.: Library of Congress, 1973), p. 29.

7. William M. Kelso, *Kingsmill Plantation, 1619–1800: Archaeology of Country Life in Colonial Virginia* (New York: Academic Press, 1984).

Index